Study Guide
for
Campbell Essential Biology

FOURTH EDITION

AND

Campbell Essential Biology with Physiology

THIRD EDITION

Study Guide
for
Campbell Essential Biology

FOURTH EDITION

AND

Campbell Essential Biology with Physiology

THIRD EDITION

Simon • Reece • Dickey

Edward J. Zalisko, Ph.D.
Blackburn College

Benjamin Cummings
Boston Columbus Indianapolis New York San Francisco Upper Saddle River
Amsterdam Cape Town Dubai London Madrid Milan Munich Paris Montréal Toronto
Delhi Mexico City São Paulo Sydney Hong Kong Seoul Singapore Taipei Tokyo

Vice President/Editor-in-Chief: Beth Wilbur
Executive Editor: Chalon Bridges
Senior Supplements Project Editor: Susan Berge
Editorial Assistant: Joshua Taylor
Managing Editor, Production: Mike Early
Production Supervisor: Jane Brundage
Production Management & Composition: S4Carlisle Publishing Services, Lori Bradshaw
Cover Production: Seventeenth Street Studios
Manufacturing Buyer: Michael Penne
Executive Marketing Manager: Lauren Harp
Senior Production Supervisor (Media): Liz Winer
Text & Cover Printer: Edwards Brothers
Cover Photos Credit: Frank Greenaway/Dorling Kindersley Media Library

ISBN 10: 0-321-64253-8
ISBN 13: 978-0-321-64253-0

Benjamin Cummings
is an imprint of

www.pearsonhighered.com

2 3 4 5 6 7 8 9 10—EB—13 12 11 10

Contents

Preface

As you begin . . .

Have you ever felt that you studied for an exam, but your grade was still low? Have you ever wished there was a better way to study?

Ask yourself these questions:

- How do you know what to study?
- How do you know when to study?
- How do you know what you already know?
- How do you know when you are done?

If these questions leave you wondering, you might benefit from the study tips in the following pages.

Ten Tips for Studying Biology

There are many subjects and many ways to study. These are tips that have proven successful for my biology students. Try them, refine them, and tailor them to your specific needs.

1. **Read your textbook assignments before lectures over the same material.** Your comprehension of the material and the quality of your notes are greatly improved if you read ahead. Think about it—if you have read ahead, the lecture material will seem familiar, your notes will make more sense, and you will need to take fewer notes because you know what is included in the textbook. *Research reveals that reading ahead is one of the most important ways to improve a student's comprehension and course grades.*

2. **Study on a regular schedule.** Waiting until the last minute to study usually produces disappointing results. One bad grade can really pull down an average of many good grades. Students who play sports, play an instrument, or have participated in a play all know about the need for *regular practice.* We all know that daily practice is required to perform well. The same is true in the classroom.
   - Start studying when the course begins. Don't wait until a week before an exam.
   - Establish good study habits early. Don't wait until you've received poor grades.
   - Know that a little bit of studying over a long period of time is much better than the same amount of studying over a short period of time. An hour a day for 15 days is many times better than 15 hours at one time (cramming). You can spend less time than cramming and get better grades!

3. **Take frequent and short study breaks.**
   - Many times when we study, our minds drift as we listen to someone in the hallway, a sound outside, or get distracted by people around us. There are two main methods to address this problem.
   - First, study in short bursts of 20 to 30 minutes each. Then take a 5–10 minute break to rest your mind. After the break, settle back in and focus intensely on your work. Repeat as necessary!
   - Second, try to study in ways that are more interactive. The additional tips below describe a highly successful interactive way to study.

4. **Identify all of the material that is addressed by the exam. Know what you need to know.**
   - After the first day of class, start identifying material for the first exam.
   - Use your course syllabus to help you determine this material.
   - If you are still unclear, talk to your instructor.

5. **Try to write many questions that address all of the information for the exam.**
   - Write questions that can be answered with short answers, 5 to 20 words or so.
   - Many students like to use 3 × 5 cards, with a question on the front and the answer on the back.

- Consider using some of the questions from the study guide.
- Begin with definitions and short lists from the material you need to study.
- Write these questions after every lecture, keeping up with the material.
- Writing the questions checks your understanding of the material. Use this opportunity to clear up confusion.

6. **Quiz yourself every day or so on all of the questions you have created.**
   - As you try to answer the questions, create a pile of cards with questions you got correct and a pile for those questions that you missed.
   - After quizzing yourself over all the questions, return to the pile of questions that you missed.
   - As you quiz yourself over the questions you missed, again create a pile of those you got correct and those questions you have now missed twice.
   - Look at the questions that you have missed twice. Use the Web activities and your textbook to review this material again. Try to figure out ways to remember this difficult information.
   - With every new lecture, add new questions to the list of questions you have already created.
7. **Stop studying when you have correctly answered every question.**
   - Finally, a way to study that builds confidence, takes less time, and has a clear end!
   - As you continue to quiz yourself, the questions will seem easier and you should feel more confident.
   - Remember to keep up with your writing and quizzing. This process doesn't work well unless done regularly.
8. **Use short 10- to 30-minute periods of your day to quiz yourself.**
   - Students that manage their time well will use the short periods before, between, and after classes to review.
   - The result is more free time.
9. **Study during the times of the day when you are most alert.**
   - Many students study in the evenings, when their mind and/or body is tired.
   - Find times in your day when you are most alert and make these your study times.
   - Use times when you tend to get sleepy for other activities (for many of us, this is early afternoon).
10. **Use the night before a test to quickly review. Then get plenty of sleep.**
    - These techniques build confidence in your knowledge of the course material.
    - Many students need only a quick review of their stack of note cards the night before the test. This might take an hour or so.
    - Then get plenty of sleep. A well-rested mind always functions better.
    - Review the questions that you missed on the exam and see if the information was in your note cards. Try to improve your study methods based upon the reasons for missing these questions.

Using Your Study Guide

The purpose of this study guide is to help you organize and review the information presented in the *Campbell Essential Biology* text. Each textbook chapter corresponds to a study guide chapter of the same number. The common components of each study guide chapter are described below in the sequence that they appear. Take a few minutes to review how they can help you master the textbook information.

Studying Advice

These study tips provide specific advice for best approaching each chapter.

Organizing Tables

These tables provide the framework for organizing complex information presented in the textbook. They help you organize the information for easier review.

Content Quiz

This is the largest section of each chapter. Some students use these questions to develop their note card questions. Others prefer to quiz themselves after they have mastered their note cards. The answers to the Content Quiz are located in the back of the study guide. The Content Quiz includes the following types of questions:

Matching—Most chapters have at least one set of matching questions which include long lists of terms or names.

Multiple Choice—These are generally short questions with one correct answer. Consider using the correct statements in these questions for note card questions.

Correctable True/False—These true/false statements have underlined words that are to be corrected if the statement is false. These corrected statements also make good material for note card questions.

Fill-in-the-Blank—These questions check your knowledge of definitions. They also make excellent note card questions.

Figure Quiz—These questions address the information presented in the textbook figures. You will need to refer to the figures in your textbook to answer them.

Analogy Questions—These questions ask you to identify relationships that are analogous to something in biology. For example, the shape of a DNA molecule is most like the shape of a spiral staircase.

Word Roots

This list presents the meanings of the word roots that form the key terms. Learning the meaning of word roots helps you guess the meaning of new terms. For example, you might know that "milli" means one thousandth and "centi" means one hundredth because you know that a millimeter is one thousandth of a meter and a centimeter is one hundredth of a meter. Therefore, you might guess that a millipede has more legs than a centipede!

Key Terms

This list consists of all the boldfaced terms and phrases in each chapter.

Crossword Puzzle

The crossword puzzle consists of the key terms and key phrases from the chapter. This is a fun way to review your knowledge of these definitions. The crossword answers are also listed at the back of the study guide.

CHAPTER 1

Introduction: Biology Today

Studying Advice

a. This introductory chapter sets the stage for the rest of the book. Here the authors focus on the concepts and less so on the terminology, so few boldfaced terms are included in this chapter.

b. If you have not already done so, read the introductory pages of this study guide to familiarize yourself with the typical components of each study guide chapter and review the advice for studying biology. This advice has proven important to thousands of my students in many different types of science classes. It might also be quite helpful to you.

c. In addition to using this study guide, spend time reviewing the media activities that are noted in your textbook and that appear on the textbook's website. These activities illustrate and review significant principles addressed in your textbook.

d. Some of the multiple-choice questions included throughout this study guide rely upon a format that you might not have often encountered before. Instead of considering a long list of incorrect statements and finding the one that is correct, you are presented with a list of true statements and asked to identify the one choice that is false. (For an example of this type of format, see question 9 in the "Content Quiz" section.) This format is increasingly included on many types of exams, so it is helpful to practice a strategy for this style. As you consider your response to this form of multiple-choice question, consider treating each choice as a true/false question. You might even note a T or F in front of the letter identifying the choice. Your final answer will then be the one statement you determine is false. In this study guide, this multiple-choice question format has the additional advantage of presenting a long list of true statements that can continue to be reviewed later. If you correct the single false statement, you add to this list of information for review. These true statements also help break the information down for the construction of note card questions.

Student Media

Activities

The Levels of Life Card Game

Energy Flow and Chemical Cycling

Classification Schemes
Darwin and the Galápagos Islands
Science, Technology, and Society: DDT

Graph It

An Introduction to Graphing

MP3 Tutors

The Process of Science

Process of Science

How Do Environmental Changes Affect a Population?
How Does Acid Precipitation Affect Trees?

Videos

Discovery Channel Video: Antibiotics
Seahorse Camouflage

You Decide

What Can We Do About Antibiotic-resistant Bacteria?

Organizing Tables

TABLE 1.1 Distinguish between prokaryotic and eukaryotic cells.

Characteristics	Prokaryotic Cells	Eukaryotic Cells
Compare the relative size of organisms in each group.		
Indicate which groups have cells that are subdivided by internal membranes forming organelles.		
List examples of each group.		

TABLE 1.2 Identify the major domains of life.

List the three domains of life.			
Indicate if the members of the group have prokaryotic cells or eukaryotic cells.			

TABLE 1.3 Compare the groups within the domain Eukarya.

	Protists	Plantae	Animalia	Fungi
Indicate if the organisms are typically unicellular or multicellular.				
Describe their mode of nutrition.				
Indicate examples of each group.				

Content Quiz

Directions: Identify the *one* best answer for the multiple-choice questions. For true/false questions, determine if the statement is true or false. If false, change the underlined word(s) to make the statement true. Finally, add the correct word(s) to the fill-in-the-blank questions to make the statements true.

Biology and Society: Biology All Around Us, The Scope of Life

THE PROPERTIES OF LIFE

1. Which one of the following is *not* a property or process of life?
   A. reproduction
   B. a tendency toward disorder
   C. growth and development
   D. energy utilization
   E. evolution

2. Which one of the following is a property of life that is best illustrated by shivering and sweating?
   A. evolution
   B. order
   C. reproduction
   D. growth and development
   E. regulation

3. True or False? Information carried by <u>proteins</u>, the units of inheritance, controls the patterns of growth and development.

4. All organisms respond to environmental ____________.

LIFE AT ITS MANY LEVELS

5. Which one of the following is a correct sequence of levels of biological organization from most general to more specific?
   A. population, organism, organ system, organs, tissues
   B. organs, cells, tissues, molecules, atoms
   C. ecosystem, population, community, organism, organ system
   D. community, population, organ system, organism, organs
   E. organism, organ system, organs, cells, tissues

6. True or False? A <u>community</u> is a group of interacting individuals of just one species.

7. Tissues are made of ____________, and molecules are made of ____________.

8. The dynamics of any ecosystem depend on two main processes. These are:
   A. the cycling of nutrients and the flow of energy from sunlight to producers and then to consumers and decomposers.
   B. the cycling of nutrients and the movement of water through the water cycle.
   C. the movement of water through the water cycle and photosynthesis.
   D. photosynthesis and cellular respiration.

9. Which one of the following statements comparing prokaryotic and eukaryotic cells is *false*?
   A. Eukaryotic cells are usually larger than prokaryotic cells.
   B. Eukaryotic cells are usually more complex than prokaryotic cells.
   C. Only eukaryotic cells have DNA; prokaryotes use RNA.
   D. Eukaryotic cells occur in animals and plants; bacteria have prokaryotic cells.

10. True or False? A <u>cell</u> is the lowest level of structure that can perform all activities required for life, including the capacity to reproduce.

11. The ____________ consists of all the environments on Earth that support life.

12. The interactions between organisms and their environments take place within ____________.

13. The ____________ is the lowest level of biological structure that can perform all activities of life.

14. The entire "book" of genetic instructions that an organism inherits is called its ____________.

LIFE IN ITS DIVERSE FORMS

Matching: Match the group on the left to its best description on the right.

_____ 15. protists

_____ 16. plants

_____ 17. fungi

_____ 18. animals

A. multicellular eukaryotes that obtain food by ingestion

B. eukaryotic organisms that are typically single-celled

C. multicellular eukaryotes that absorb nutrients by breaking down dead organisms and organic wastes

D. multicellular eukaryotes that produce their own sugars and other foods by photosynthesis

19. Which one of the following choices correctly lists the groups in their order of abundance, from the group with the greatest number of species to the group with the fewest species?
    A. plants, insects, vertebrates
    B. vertebrates, plants, insects
    C. plants, vertebrates, insects
    D. insects, plants, vertebrates
    E. insects, vertebrates, plants

20. Which of the following are two domains that include organisms with prokaryotic cells?
    A. bacteria and eukarya
    B. bacteria and archaea
    C. archaea and eukarya
    D. bacteria and fungi

21. Look at Figure 1.8 of your text. Which one of the following groups contains the smallest organisms?
    A. archaea
    B. protists
    C. plantae
    D. animalia
    E. fungi

22. True or False? All life is presently classified into <u>four</u> domains.

23. The branch of biology that names and classifies species is called ____________.

24. The kingdoms of life can be assigned to even higher levels of classification called ____________.

Evolution: Biology's Unifying Theme

THE DARWINIAN VIEW OF LIFE

25. Which one of the following traits most likely occurred in the most recent common ancestor of a lizard, pigeon, turtle, and bear?
    A. fur
    B. feathers
    C. shell
    D. wings
    E. backbone

26. In his book *The Origin of Species*, Charles Darwin developed two main points. These were that:
    A. the world is very old and evolution occurs by natural selection.
    B. evolution occurs slowly and new species form by spontaneous generation.
    C. new species evolve by mutations and new species don't reproduce with old species.
    D. modern species descended from ancestral species and organisms evolve by natural selection.

27. Examine the relationships between bears in the diagram in Figure 1.10 of your text. Which one of the following bears is most closely related to the polar bear?
    A. giant panda
    B. brown bear
    C. spectacled bear
    D. sun bear
    E. sloth bear

28. True or False? The common ancestor of bears and chipmunks <u>had</u> hair and mammary glands.

29. ____________ is the theme that unifies all of biology.

30. In *The Origin of Species*, Darwin proposed a mechanism for descent with modification, which he called ____________.

NATURAL SELECTION

31. Which one of the following pairs of facts was the basis for Darwin's conclusion of unequal reproductive success?
    - A. individual variation; offspring are the products of a blending of the parental genetics
    - B. individual variation; overproduction and competition
    - C. overproduction and competition; males usually fight for a mate
    - D. parents produce offspring similar to themselves; overproduction and struggle for existence
    - E. males usually fight for a mate; parents produce offspring similar to themselves

32. Darwin could see that in artificial selection, humans are substituting for:
    - A. individual variation.
    - B. overproduction of offspring.
    - C. the generation of genetic diversity.
    - D. the environment.

33. In the example of the beetles discussed in the text:
    - A. the most common beetle color changed after the environment changed.
    - B. the environment created favorable characteristics.
    - C. natural selection favored traits that worked best in different environments.
    - D. the lightest beetle color pattern was most adaptive.

34. What do artificial selection and natural selection have in common?
    - A. Both result from a need or desire.
    - B. Both are purposeful processes.
    - C. Both rely upon individual variation.
    - D. Both are directional, with selection toward a goal.

35. Evolution works most like:
    - A. remodeling an old home.
    - B. an architect designing a new home.
    - C. people in a community planning where to build a park.
    - D. people voting in an election.

36. True or False? The product of natural selection is <u>adaptation</u>.

37. Darwin used the phrase ____________ to refer to unequal reproductive success.

The Process of Science

DISCOVERY SCIENCE

38. Which one of the following is an example of discovery science?
    A. comparing the effects of two different drugs on the healing of wounds
    B. testing to see if large doses of vitamin C will prevent the common cold
    C. describing the structure of a newly discovered dinosaur skull
    D. testing tropical plants for chemicals that might help fight cancer

39. True or False? An ecologist describing all the animals and plants in a region of a tropical rain forest is using <u>hypothesis-driven</u> science.

40. The word ____________ is derived from a Latin verb meaning "to know."

41. Discovery science relies upon ____________ reasoning, generalizing from many observations.

HYPOTHESIS-DRIVEN SCIENCE

42. Consider the following statement. "If all vertebrates have backbones, and turtles are vertebrates, then turtles have backbones." This statement is an example of:
    A. a hypothesis.
    B. discovery science logic.
    C. rationalization.
    D. deductive reasoning.

43. True or False? <u>Inductive</u> reasoning flows from the general to the specific.

44. A tentative answer to a question defines a(n) ____________.

THE PROCESS OF SCIENCE: IS TRANS FAT BAD FOR YOU?

45. In the trans fat study investigating the kinds of fat in people who had a heart attack, the patients who did not have a heart attack represented the:
    A. control group.
    B. experimental group.
    C. variable.
    D. hypothesis.
    E. observation.

46. The study of over 120,000 female U.S. nurses was an example of ____________ science.

47. In most cases, a control group and an experimental group differ by just one ____________.

THEORIES IN SCIENCE

48. Which one of the following does *not* apply to scientific theories? Scientific theories:
    A. provide a comprehensive explanation.
    B. are another name for a hypothesis.
    C. are supported by abundant evidence.
    D. are widely accepted.

49. The one theme that continues to hold all of biology together is ____________.

THE CULTURE OF SCIENCE; SCIENCE, TECHNOLOGY, AND SOCIETY

50. Which one of the following is *not* a characteristic of modern science?
    A. repeatability of experiments
    B. dependence upon observations
    C. requirement that ideas be testable
    D. independence and isolation of researchers

51. True or False? Scientists usually pay <u>little</u> attention to other researchers currently working on the same problem.

52. Science and ____________ are interdependent.

Evolution Connection: Evolution in Our Everyday Lives

53. Which one of the following is the best description of how antibiotic-resistant bacteria evolve?
    A. Antibiotics cause bacteria to change and become resistant.
    B. Bacteria need to be resistant to antibiotics so the bacteria change.
    C. Bacteria that are naturally resistant are favored by the use of antibiotics.
    D. When antibiotics are used, bacteria need to survive and therefore must change or die.

54. Which one of the following environmental factors contributes most to the evolution of antibiotic-resistant bacteria?
    A. mutations from ultraviolet light
    B. the use of antibiotics
    C. the need for bacteria to change
    D. the evolution of viruses

Word Roots

eco = house (ecology: the study of how organisms interact with their environments)
hypo = below (hypothesis: a tentative explanation)

Key Terms

biology
case study
controlled experiment
discovery science
ecosystem
hypothesis
hypothesis-driven science
life
natural selection
scientific method
theory

Crossword Puzzle

Use the Key Terms list from this chapter to fill in the crossword puzzle.

ACROSS

3. a series of steps that guide scientific investigations
6. a biological community and its environment
7. a type of science that includes observations, measurements, and other descriptions
8. a widely accepted explanatory idea supported by a large body of evidence
9. a phenomenon that includes order, regulation, growth, development, reproduction, energy utilization, response, and evolution
10. a type of experiment designed to compare an experimental group with a control group
11. the type of science that tests tentative explanations

DOWN

1. the scientific study of life
2. a way that a population of organisms can change over generations because of inheritable differences in reproduction
4. an in-depth examination of an actual investigation
5. a tentative explanation

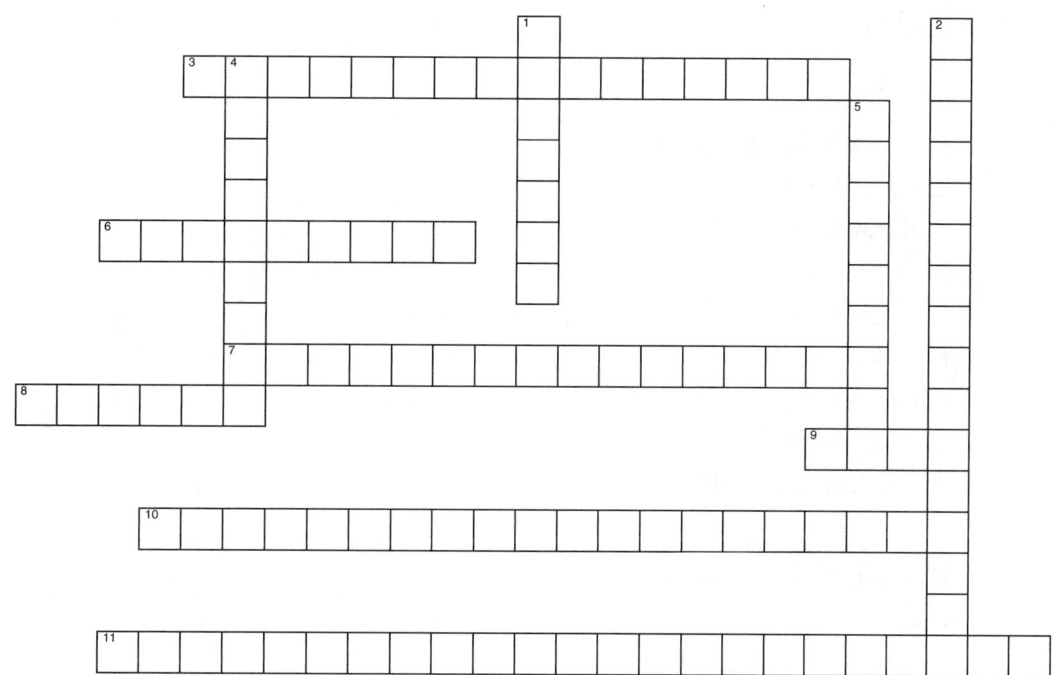

CHAPTER 2

Essential Chemistry for Biology

Studying Advice

a. If this is your first experience with chemistry, work slowly to master the terminology and basic relationships of atoms and their interactions. Before reading the chapter, look through the figures and read the chapter summary. This will give you an idea of the subjects that will be addressed in the text.

b. Understanding the terminology and ideas in this chapter is essential to understanding the content in the next few chapters. These next chapters continue to explore the molecules of life and how they interact to form cells. Your investment of energy in this chapter will pay dividends in the chapters to come!

c. Finally, think carefully about *how* you study. Refer to the study advice at the start of this study guide and establish good habits now.

Student Media

Activities

Structure of the Atomic Nucleus

Electron Arrangement

Build an Atom

Ionic Bonds

Covalent Bonds

Hydrogen Bonds

Polarity of Water

Cohesion of Water

Acids, Bases, and pH

BLAST Animations

Covalent Bonds

Hydrogen Bonds in Water

MP3 Tutors

The Properties of Water

Process of Science

How Are Space Rocks Analyzed for Signs of Life?
How Does Acid Precipitation Affect Trees?

Organizing Tables

TABLE 2.1 Compare the three main subatomic particles.

	Relative Size Compared to Other Parts of an Atom	Electrical Charge, if Any	Location in an Atom
Proton			
Neutron			
Electron			

TABLE 2.2 Compare the three main types of chemical bonds.

	Nature of the Bond	Example
Ionic bond		
Covalent bond		
Hydrogen bond		

TABLE 2.3 Describe your own examples for each of the properties of water.

Property of Water	Examples from Your Life
The cohesion of water and surface tension	
How water moderates temperature	
Ice floating	
Water as a solvent	

Content Quiz

Directions: Identify the *one* best answer for the multiple-choice questions. For true/false questions, determine if the statement is true or false. If false, change the underlined word(s) to make the statement true. Finally, add the correct word(s) to the fill-in-the-blank questions to make the statements true.

Biology and Society: Fluoride in the Water

1. Fluoride prevents cavities by:
   A. killing viruses in the mouth.
   B. promoting the production of new teeth.
   C. promoting the replacement of lost minerals on the tooth surface.
   D. encouraging the growth of special bacteria that add to the tooth surface.
   E. None of the above.

2. True or False? Fluoride is <u>a common</u> ingredient in Earth's crust.

3. A sticky layer of bacteria, dead cells, food, and saliva that coats your teeth is called ____________.

Some Basic Chemistry

MATTER: ELEMENTS AND COMPOUNDS

4. Four elements make up 96% of the human body. Which one of the following is *not* one of those four major elements?
   A. carbon
   B. nitrogen
   C. iron
   D. oxygen
   E. hydrogen

5. True or False? Carbon dioxide is an example of a <u>compound</u>, because it contains two or more elements in a fixed ratio.

6. Elements that your body needs in very small amounts are called ____________.

ATOMS

7. Which one of the following statements about atoms is *false*?
   A. All atoms of an element have the same number of protons.
   B. Atoms whose outer shells are not full tend to interact with other atoms.
   C. An atom is the smallest unit of matter that retains the properties of an element.
   D. A proton and an electron are almost identical in mass.
   E. The farther an electron is from the nucleus, the greater its energy.

8. The way that a satellite orbits Earth is most like the relationship between:
   A. a proton and a neutron of an atom.
   B. a cell and its molecules.
   C. an electron and the nucleus of an atom.
   D. organisms in an ecosystem.
   E. a cell in a tissue.

9. Examine Figure 2.5 of your text, which includes a helium atom. Which one of the following is a problem with the way this helium atom is represented?
   A. The protons and neutrons do not form the nucleus.
   B. Electrons are actually larger than protons and neutrons.
   C. Electrons do not orbit around the nucleus.
   D. The distance between the electrons and the nucleus is many times greater.
   E. Sometimes the electrons contribute to the structure of the nucleus.

10. True or False? Isotopes of an element have the same number of <u>protons</u> and <u>electrons</u> but different numbers of <u>neutrons</u>.

11. The mass number is the sum of the number of _____________ and ______________.

CHEMICAL BONDING AND MOLECULES

12. When a covalent bond occurs,
    A. protons are transferred from one atom to another.
    B. the atoms in the reaction become positively charged.
    C. an atom gives up one or more electrons to another atom.
    D. ions are formed.
    E. two atoms share one or more electrons.

13. A person borrowing money from a bank is most like the relationship between:
    A. atoms that form a covalent bond.
    B. atoms that form an ionic bond.
    C. a proton and a neutron in an atom.
    D. two isotopes of an element.
    E. an organism and its cells.

14. Examine the electrons in the atoms in Figure 2.7 of your text. How many atoms of oxygen will covalently bond to one atom of carbon?
    A. one atom of oxygen
    B. two atoms of oxygen
    C. three atoms of oxygen
    D. four atoms of oxygen
    E. Oxygen will not covalently bond to carbon.

15. Water molecules are attracted to each other because of:
    A. covalent bonds.
    B. hydrogen bonds.
    C. atomic bonds.
    D. ionic bonds.

16. When we get out of a shower or bath, the surface of our skin is mostly wet because:
    A. water is being released by our skin cells.
    B. water is a polarized molecule that sticks to surfaces.
    C. water from the air condenses on our skin.
    D. our bodies continue to produce water.

17. True or False? In a water molecule, the electrons spend most of their time near the <u>hydrogen atoms</u>.

18. True or False? A molecule of carbon dioxide is a <u>compound</u> because it consists of two or more elements.

19. Water's two hydrogen atoms are joined to the oxygen atom by a(n) ____________ bond.

20. Atoms that are electrically charged because of gaining or losing electrons are called ____________.

CHEMICAL REACTIONS

21. In the chemical reaction of $2\ H_2 + O_2 \rightarrow 2\ H_2O$,
    A. water is a reactant.
    B. hydrogen and oxygen are the products.
    C. the number of atoms of reactants is more than the number of atoms of the products.
    D. a chemical compound is formed.

22. True or False? Chemical reactions <u>cannot</u> destroy or create matter.

23. New chemical bonds are formed in the ____________ of a reaction.

Water and Life

WATER'S LIFE-SUPPORTING PROPERTIES

24. Which one of the following statements about water is *false*?
    A. Covalent bonds give water unusually high surface tension.
    B. Water absorbs and stores a large amount of heat while warming up only a few degrees.
    C. Evaporative cooling occurs because water molecules with the greatest energy vaporize first.
    D. Ice floats because water molecules move farther apart when they form a solid.
    E. Water is a common solvent inside your body.

25. Which one of the following is the best example of surface tension?
   A. floating ice
   B. a water strider "standing" on the surface of liquid water
   C. sweat cooling the surface of your skin
   D. sugar dissolving quickly in a cup of hot tea
   E. a squirrel standing on an ice-covered surface of a pond

26. It is a hot summer day and Jill just finished jogging. As she grabs a drink, ice cubes floating at the top of the glass hit her nose and lemonade spills out. She uses her towel to wipe up drops of lemonade clinging to her cheeks. Which one of the following properties of water was *not* described in this situation?
   A. evaporative cooling
   B. the cohesive properties of water
   C. the ability of ice to float
   D. All of the above were noted in the description.

27. Which one of the following is represented in Figure 2.11 of your text?
   A. surface tension
   B. ability of water to moderate temperature
   C. ability of ice to float
   D. evaporative cooling
   E. versatility of water as a solvent

28. Your instructor calculates the average score on the last exam. Then your instructor eliminates the top three scores and recalculates the average, which is now quite a bit lower. This recalculation is most like the way that:
   A. sweat cools your skin through evaporative cooling.
   B. ice is able to expand, permitting it to float.
   C. molecules of water release energy to form surface tension.
   D. solutes are broken down when dissolved in a solvent.

29. True or False? Because of <u>hydrogen bonding</u>, water has a better ability to resist temperature change than most other substances on Earth.

30. A(n) ____________ is dissolved in a solvent to produce a(n) ____________.

31. When ____________ is the solvent, the result is an aqueous solution.

THE PROCESS OF SCIENCE: CAN EXERCISE BOOST YOUR BRAIN POWER?

32. The experiment examining the relationships between brain size and exercise concluded that aerobic exercise:
    A. can increase brain size.
    B. can decrease brain size.
    C. has no effect on brain size.

33. True or False? MRI depends on the behavior of <u>carbon</u> atoms in water molecules.

34. In the experiment examining the relationships between brain size and exercise, a group of 29 subjects engaged in nonaerobic exercise served as the ____________ group.

ACIDS, BASES, AND PH

35. The pH of a solution:
    A. inside most living cells is close to 5.
    B. can be quickly changed by the addition of a buffer.
    C. can range from 0 to 14.
    D. is acidic above the level of 7.
    E. is determined by the relative amount of OH^-.

36. A solution with a pH of 7 contains:
    A. more H^+ than OH^-.
    B. more OH^+ than H^-.
    C. only H^+.
    D. only OH^-.
    E. equal amounts of H^+ and OH^-.

37. Which one of the following people functions most like a buffer?
    A. a comedian who keeps the audience laughing
    B. a funeral director who helps people in times of grief
    C. a counselor who tries to understand the hidden meaning in your dreams
    D. a coach pumping up a losing team and challenging a winning team
    E. an actor playing a serious role in a drama

38. True or False? A solution with a pH of 6 has <u>100 times</u> more H^+ than an equal amount of a solution with a pH of 5.

39. Most biological fluids contain ____________, substances that resist changes in pH.

40. A chemical compound that releases hydrogen ions in a solution is called a(n) ____________. A chemical compound that accepts hydrogen ions and removes them from a solution is called a(n) ____________.

Evolution Connection: The Search for Extraterrestrial Life

41. We expect that if life similar to ours evolved elsewhere in the universe, it too would depend on:
    A. minerals.
    B. oxygen.
    C. carbon dioxide.
    D. water.
    E. hydrogen.

42. True or False? Recent NASA missions to Mars suggest that water was once abundant on Mars.

Word Roots

aqua = water (aqueous: a type of solution in which water is the solvent)
co = together; **valent** = strength (covalent bond: an attraction between atoms that share one or more pairs of outer-shell electrons)
iso = equal (isotope: an element having the same number of protons and electrons but a different number of neutrons)
neutr = neither (neutron: a subatomic particle with a neutral electrical charge)

Key Terms

acid	compound	isotope	product
aqueous solution	covalent bond	mass	proton
atom	electron	mass number	radioactive isotope
atomic number	element	matter	reactant
base	evaporative cooling	molecule	solute
buffer	heat	neutron	solution
chemical bond	hydrogen bond	nucleus	solvent
chemical reaction	ion	pH scale	temperature
cohesion	ionic bond	polar molecule	trace element

Crossword Puzzle

Use the Key Terms list from this chapter to fill in the crossword puzzle.

ACROSS

1. a process leading to chemical changes in matter
4. an atom's central core
5. a mixture of two or more substances, one of which is water
7. an attraction between two ions with opposite electrical charges
9. a substance that decreases the hydrogen concentration in a solution
11. a measure of the amount of material in an object
12. a substance that cannot be broken down into other substances
13. a group of two or more atoms held together by covalent bonds
14. a chemical substance that resists changes in pH
15. a subatomic particle with a single positive electrical charge
17. a subatomic particle that is electrically neutral
18. the smallest unit of matter that retains the properties of an element
20. anything that occupies space and has mass
21. an attraction between atoms that share one or more pairs of electrons
22. a measure of the intensity of heat
23. a substance that increases the hydrogen ion concentration in a solution
25. an element that is essential for the survival of an organism but only in minute quantities
26. the attraction between molecules of the same kind
27. the dissolving agent in a solution
29. the number of protons in each atom of a particular element
30. a molecule that has opposite charges on opposite ends
31. a fluid mixture of two or more substances
32. a type of isotope whose nucleus decays spontaneously

DOWN

1. an attraction between two atoms
2. the sum of the number of protons and neutrons in an atom's nucleus
3. a measure of the relative acidity of a solution, ranging from a value of 0 to 14
6. surface cooling that results when a substance evaporates
7. an atom or molecule that has gained or lost one or more electrons
8. a variant form of an atom
10. the amount of energy associated with the movement of the atoms and molecules in a body of matter
12. a subatomic particle with a negative charge
16. a starting material in a chemical reaction
19. a weak chemical bond between a partially positive hydrogen and a partially negative atom
24. a substance containing two or more elements in a fixed ratio
27. a substance that is dissolved in a solution
28. an ending material in a chemical reaction

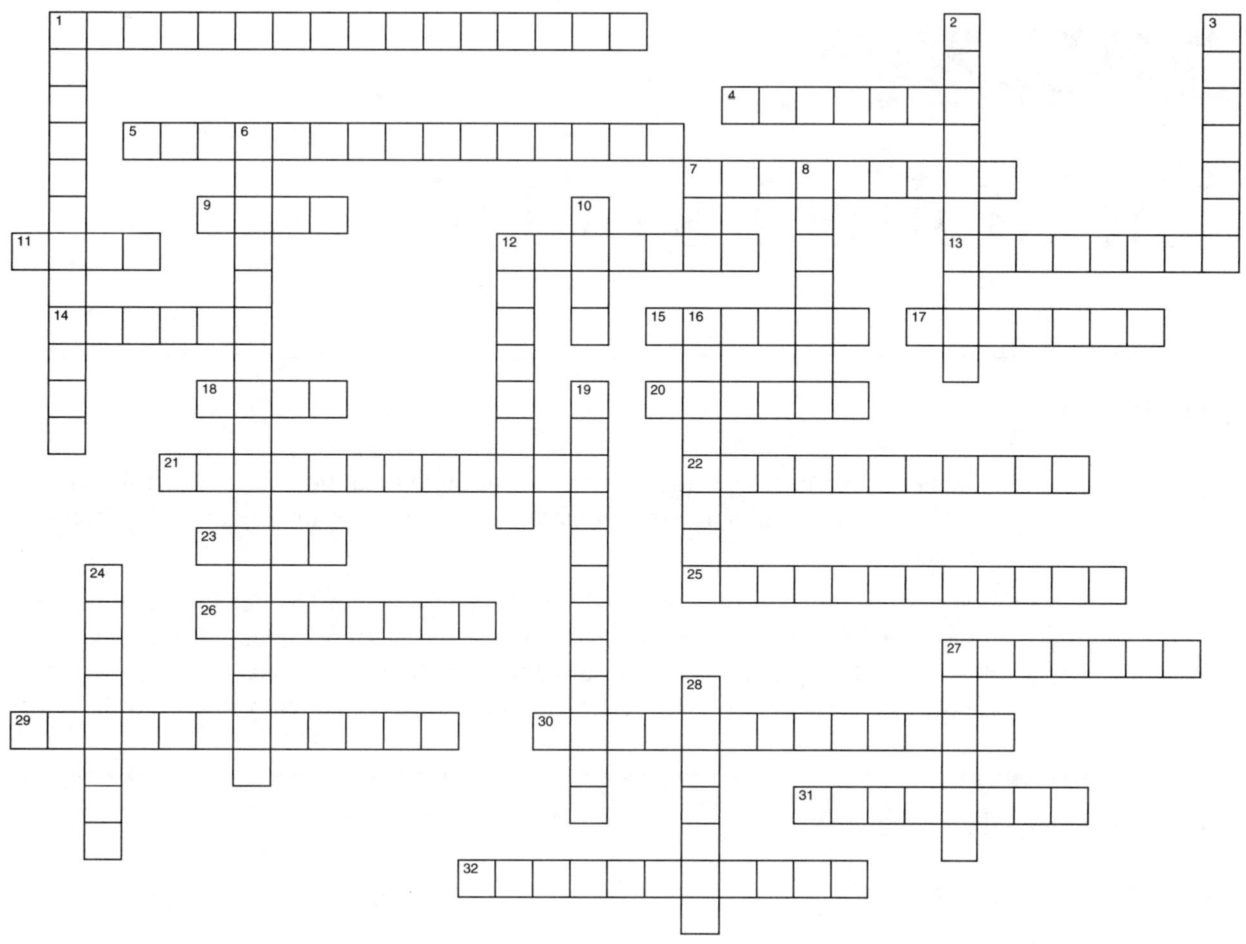

CHAPTER 3

The Molecules of Life

Studying Advice

a. Chapter 3 introduces the four main types of biological molecules. The following organizing table will help you compare these categories and organize the details of each group.

b. Find a nutrition label on some packaged food around you. Notice that three of the four categories of biological molecules discussed in this chapter are sources of nutrition. These three types—carbohydrates, proteins, and lipids—are all sources of calories. The chemistry in this chapter is something you already know a little about.

Student Media

Activities

Diversity of Carbon-Based Molecules
Functional Groups
Making and Breaking Polymers
Models of Glucose
Carbohydrates
Lipids
Protein Functions
Protein Structure
Nucleic Acid Functions
Nucleic Acid Structure

Biology Labs On-Line

HemoglobinLab

BLAST Animations

Alpha Helix
Protein Primary Structure
Protein Secondary Structure
Protein Tertiary and Quaternary Structure

MP3 Tutors

Protein Structure and Function
DNA Structure

Process of Science

What Factors Determine the Effectiveness of Drugs?

You Decide

Low-Fat or Low-Carb Diets—Which is Healthier?

Organizing Tables

TABLE 3.1 Compare the four classes of large biological molecules by completing this table. Some of the cells are already filled in to help you complete the table.

Group	Monomer	Subgroups	Examples
Carbohydrates		Monosaccharides	
		Disaccharides	
		Polysaccharides	
Lipids	*No Monomer*	Unsaturated fats	
		Saturated fats	
		Steroids	
Proteins		*No Subgroup*	
Nucleic acids		DNA	
		RNA	

Content Quiz

Directions: Identify the *one* best answer for the multiple-choice questions. For true/false questions, determine if the statement is true or false. If false, change the underlined word(s) to make the statement true. Finally, add the correct word(s) to the fill-in-the-blank questions to make the statements true.

Biology and Society: Got Lactose?

1. An insufficient amount of lactase results in:
   A. excessive heartburn.
   B. digestion of lactose in the large intestine, producing discomfort.
   C. accumulation of lactose in the muscles, causing muscle weakness.
   D. increased blood levels of lactose, resulting in disorientation and headaches.
   E. None of the above.

2. True or False? Lactose is commonly found in <u>milk</u>.

3. The enzyme that helps to break down lactose is called ____________.

Organic Compounds

CARBON CHEMISTRY

4. Which one of the following statements about organic molecules is *false*?
   A. It is possible to construct an endless diversity of carbon skeletons.
   B. The simplest hydrocarbon is methane.
   C. Carbon can use only one bond to attach to another carbon atom.
   D. Carbon completes its outer shell by sharing electrons with up to four other atoms.
   E. Carbon frequently bonds to the elements hydrogen, oxygen, and nitrogen.

5. Examine the bonds between the carbons in Figure 3.1 of your text. Find a pair of carbons with a double line between them. The double line indicates that these two carbons:
   A. share two electrons.
   B. share two hydrogens.
   C. are joined by an ionic bond.
   D. are joined at the nucleus.
   E. share electrons with hydrogens that we cannot see.

6. True or False? The groups of atoms that usually participate in chemical reactions are called <u>functional groups</u>.

7. In methane, carbon is joined to four hydrogen atoms by ____________ bonds.

GIANT MOLECULES FROM SMALLER BUILDING BLOCKS

8. Which one of the following processes is the reverse of a dehydration reaction?
   A. hydrolysis
   B. hydration
   C. denaturation
   D. diffusion
   E. osmosis

9. True or False? In the process of a dehydration reaction, a molecule of <u>carbon dioxide</u> is formed.

10. In the same way that a train is made by linking together railroad cars, cells make _____________ by linking together _____________.

Large Biological Molecules

CARBOHYDRATES

11. Automobiles use gasoline the way that cells use:
    A. cellulose.
    B. amino acids.
    C. glucose.
    D. proteins.
    E. lipids.

12. Sucrose is:
    A. the main carbohydrate in plant sap.
    B. the main sweetener in soft drinks.
    C. a monosaccharide.
    D. sweeter tasting than fructose.
    E. abundant in corn syrup.

13. Plants use starch the way that animals use:
    A. glucose.
    B. glycogen.
    C. fructose.
    D. cellulose.
    E. sucrose.

14. Compare the structures of glucose and fructose in Figure 3.5 of your text. How are these molecules different?
    A. Glucose has more double bonds.
    B. Fructose has more carbon atoms.
    C. Fructose has more hydrogen atoms.
    D. Glucose has more oxygen atoms.
    E. Only the shapes are different.

15. Examine the structure of glucose in Figure 3.6 of your text. How are the linear and ring structures different?
    A. The linear structure has more carbon atoms.
    B. The linear structure has more oxygen atoms.
    C. The ring structure has more hydrogen atoms.
    D. The ring structure has more double bonds.
    E. Only the shapes are different.

16. True or False? <u>Starch</u> is the most abundant organic compound on Earth.

17. True or False? Cellulose <u>cannot</u> be hydrolyzed by most animals.

18. Glucose and fructose are examples of ____________, molecules that match in their molecular formulas but differ in their shapes.

19. Cells construct a disaccharide by joining two ____________ in a(n) ____________ reaction.

20. Starch is a polysaccharide made by joining together many ____________ monomers.

LIPIDS

21. Unsaturated fats:
    A. are usually solid at room temperature.
    B. contribute to cardiovascular disease more than saturated fats.
    C. have the maximum number of hydrogen atoms attached.
    D. lack double bonds in their hydrocarbon portions.
    E. None of the above.

22. If we add hydrogen to an unsaturated fat, we would expect that the unsaturated fat would:
    A. have more double bonds.
    B. be a much longer molecule.
    C. be stiffer at room temperature.
    D. likely be healthier.
    E. None of the above.

23. Which one of the following statements about cholesterol is *false*? Cholesterol is:
    A. made of a carbon skeleton consisting of four fused rings.
    B. one of the most well-known steroids.
    C. used by our bodies to make sex hormones.
    D. similar to fats in structure and function.
    E. present in cell membranes.

24. Anabolic steroids:
    A. structurally resemble testosterone.
    B. can cause serious mental problems.
    C. are synthetic variants of testosterone.
    D. can damage the liver and lead to infertility.
    E. All of the above.

25. True or False? A pound of fat contains <u>more than twice</u> as much energy as a pound of starch.

26. The accumulation of lipid-containing deposits within the walls of blood vessels leads to a type of cardiovascular disease called ____________.

27. Lipids are ____________, which means that they cannot mix with water.

PROTEINS

28. Which one of the following statements about proteins is *false*?
    A. A protein's three-dimensional shape enables it to carry out its normal functions.
    B. All proteins are made from a common set of just 20 amino acids.
    C. Amino acids in a protein are linked together by peptide bonds.
    D. Enzymes are specialized types of proteins.
    E. Proteins are made from nucleic acid monomers.

29. Which one of the following is most like the process of making proteins?
    A. weaving hair into a braid
    B. sewing a quilt from hundreds of patches of old clothing
    C. making a train by connecting together 20 different types of railroad cars
    D. gathering peas onto a spoon as you prepare to eat them
    E. preparing a stew by cutting up four different vegetables and mixing them with meat

30. The way a protein changes shape when it is heated is most like:
    A. tearing sections off a roll of toilet paper.
    B. cutting a piece of string into two.
    C. separating a few cars from a long train.
    D. untangling a garden hose.

31. Examine Figure 3.20 of your text to best understand the levels of protein structure. Which one of the following is most like the secondary level of protein structure?
    A. the sequence of rail cars on a train
    B. a coiled spring
    C. the sequence of letters that form a word
    D. the links of chain forming a necklace

32. True or False? <u>All amino acids</u> have a carboxyl group, an amino group, and a hydrogen atom attached to a central carbon atom.

33. The specific amino acid sequence of a protein is the protein's ____________ structure.

34. If a protein is exposed to a change in temperature or pH, it can lose its normal shape in a process called ____________.

NUCLEIC ACIDS, THE PROCESS OF SCIENCE:
DOES LACTOSE INTOLERANCE HAVE A GENETIC BASIS?

35. Which one of the following statements about nucleic acids is *false*?
   A. The base is the part that varies between nucleotides of DNA.
   B. There are two types of nucleic acids, DNA and RNA.
   C. Nucleic acids are linked into long strands to form nucleotides.
   D. The sequence of amino acids in proteins is determined by the sequence of nucleotides in DNA.
   E. Molecules of DNA form a double helix.

36. The overall shape of a DNA molecule is most similar to the shape of:
   A. railroad tracks.
   B. a spiral staircase.
   C. a screwdriver.
   D. a bicycle chain.
   E. the number "8."

37. DNA and RNA are alike in that both:
   A. are usually double-stranded molecules.
   B. use the bases adenine, guanine, cytosine, and thymine.
   C. use the sugar deoxyribose.
   D. are polymers of nucleotides.
   E. None of the above.

38. True or False? The double-stranded genetic material humans and other organisms pass from one generation to the next is <u>RNA</u>.

39. True or False? Lactose intolerance results from a mutation <u>within</u> the lactase gene.

40. Nucleic acids are made of monomers called ______________.

Evolution Connection: Evolution and Lactose Intolerance in Humans

41. Lactose intolerance is most *rare* in Americans:
   A. from Africa.
   B. from Asia.
   C. native to North America.
   D. from northern Europe.

42. True or False? Lactose intolerance is <u>most</u> common in human cultures that relied upon dairy products.

Word Roots

di = two; **sacchar** = sugar (disaccharide: two monosaccharides joined together)
glyco = sweet; **gen** = producing (glycogen: a polysaccharide sugar used to store energy)
hydro = water; **lyse** = break (hydrolysis: breaking chemical bonds by adding water)
iso = equal; **meros** = part (isomer: molecules with similar molecular formulas but different structures)
macro = big (macromolecules: a giant molecule in living organisms)
mono = single (monosaccharide: simplest type of sugar)
philic = loving (hydrophilic: water-loving property of a molecule)
phobos = fearing (hydrophobic: water-repelling property of a molecule)
poly = many (polymer: a chain made from smaller organic molecules)
sclero = hard (atherosclerosis: hardening of the arteries)
tri = three (triglyceride: a glycerol molecule joined with three fatty acid molecules)

Key Terms

amino acid
atherosclerosis
carbohydrates
cellulose
dehydration reaction
denaturation
disaccharides
DNA
double helix
fat
functional groups
gene
glycogen
hydrocarbons
hydrogenation
hydrolysis
hydrophilic
hydrophobic
isomers
lipids
macromolecules
monomers
monosaccharides
nucleic acids
nucleotides
organic compounds
peptide bond
polymers
polypeptide
polysaccharides
primary structure
protein
RNA
saturated
starch
steroids
sugar-phosphate backbone
trans fat
triglyceride
unsaturated

Crossword Puzzle

Use the Key Terms list from this chapter to fill in the crossword puzzle.

ACROSS

1. the type of backbone forming polynucleotides
3. the genetic material that organisms inherit from their parents
8. a polysaccharide used to store energy in animals
9. type of organic molecule that includes sugars and starch
11. process that links monomers together
13. the addition of hydrogens to unsaturated fats
15. type of group of atoms usually involved in chemical reactions
16. the most abundant organic compound on Earth
17. molecules that have the same molecular formula but different structures
19. a glycerol molecule joined with three fatty acid molecules
20. a change in the shape of a protein
21. water-hating property of a molecule
22. shape of DNA
23. a type of organic molecule that includes fats, steroids, and phospholipids
25. two monosaccharides joined together
27. many monosaccharides joined together
28. a polysaccharide used to store energy in plants
29. type of fatty acid with the maximum number of hydrogens
31. type of fatty acid with less than the maximum hydrogens
32. type of organic molecule including DNA and RNA
33. an unhealthy unsaturated fat produced by hydrogenation
34. a chain made from smaller organic molecules
35. the type of compounds that are carbon-based
36. first level of protein structure
37. type of bond between adjacent amino acids
38. a large lipid molecule made from glycerol and three fatty acids

DOWN

1. a type of lipid with a carbon skeleton in the form of four fused rings
2. building block of a polymer
4. cardiovascular disease in which lipid deposits accumulate in walls of arteries
5. breaking chemical bonds by adding water
6. the simplest organic molecule
7. a specific stretch of DNA that directs the production of a polypeptide
10. water-loving property of a molecule
12. monomer of a protein
14. a type of nucleic acid that uses the base uracil
18. simplest type of sugar
24. a chain of amino acids used to make a protein
26. macromolecule made of one or more polypeptides
30. gigantic molecule
32. monomer of a nucleic acid

1
2
3
4
5
6
7
8
9
10
11
12
13
14
15
16
17
18
19
20
21
22
23
24
25
26
27
28
29
30
31
32
33
34
35
36
37
38

CHAPTER 4

A Tour of the Cell

Studying Advice

a. Before you begin reading this chapter, examine Figure 4.5 of your text. How many of these cell parts do you recognize from prior courses? Also, refer to Figure 4.3 of your text to become familiar with the relative sizes of cells and their parts.

b. The following organizing tables will be especially useful in this chapter. Many basic parts of cells are introduced. These tables will help you keep the information organized for review.

Student Media

Activities

Metric System Review

Prokaryotic Cell Structure and Function

Comparing Cells

Build an Animal Cell and a Plant Cell

Membrane Structure

Role of the Nucleus and Ribosomes in Protein Synthesis

The Endomembrane System

Build a Chloroplast and a Mitochondrion

Cilia and Flagella

Review: Animal Cell Structure and Function

Review: Plant Cell Structure and Function

BioFlix

Tour of an Animal Cell

Tour of a Plant Cell

BLAST Animations

Animal Cell Overview

Plant Cell Overview

Vesicle Transport along Microtubules
Vacuole
Mitochondrion
Plant Cell Wall

MP3 Tutors

Cell Organelles

Process of Science

What Is the Size and Scale of Our World?

Videos

Discovery Channel Video: Cells
Prokaryotic Flagella
Euglena
Cytoplasmic Streaming
Chlamydomonas
Paramecium Cilia
Paramecium Vacuole

Organizing Tables

TABLE 4.1 Compare the features of prokaryotic and eukaryotic cells. (See your text and Figures 4.3 and 4.5 of your text.)

	Prokaryotic Cells	Eukaryotic Cells
Which group has a nucleus bordered by a membrane?		
Which group has organelles?		
How do the sizes compare?		
What are examples of each group?		

TABLE 4.2 Compare the structure, location, and function of the components of the endomembrane system.

	Structure	Location in the Cytoplasm	Function(s)
Smooth endoplasmic reticulum			
Rough endoplasmic reticulum			
Golgi apparatus			
Lysosome			
Central vacuole			

TABLE 4.3 Compare the features of cilia and flagella.

	Cilia	Flagella
Which are usually longer?		
Which are usually more numerous?		
How does their basic architecture compare?		
Where are they found?		

TABLE 4.4 Compare the structures and functions of mitochondria and chloroplasts.

	Mitochondria	Chloroplasts
These organelles are found in the cells of organisms in what kingdoms?		
Which organelle is involved in cellular respiration and which is involved in photosynthesis?		
Compare the number of membranes and compartments found in each organelle.		
Draw a general sketch to show the locations of the membranes and compartments.		
What part of each organelle is the most active region?		

Content Quiz

Directions: Identify the *one* best answer for the multiple-choice questions. For true/false questions, determine if the statement is true or false. If false, change the underlined word(s) to make the statement true. Finally, add the correct word(s) to the fill-in-the-blank questions to make the statements true.

Biology and Society: Drugs That Target Bacterial Cells

1. Most antibiotics harm bacteria but do little damage to the human host by:
   A. killing only rapidly dividing cells.
   B. binding to structures found only in bacteria.
   C. attacking only large cells.
   D. destroying only cells that have a nucleus.

2. True or False? Penicillin was first isolated from <u>algae</u> in 1928.

3. Drugs that disable or kill infectious bacteria are called ____________.

The Microscopic World of Cells

4. Which one of the following types of microscopes would you most likely use in a laboratory to study cells on a prepared microscope slide?
   A. transmission electron microscope
   B. scanning electron microscope
   C. light microscope

5. Which one of the following is most closely associated with the term "resolution"?
   A. clarity
   B. larger size
   C. greater color
   D. lighter image

6. Examine Figure 4.5 of your text. Which one of the following organelles is physically connected to the nucleus?
   A. mitochondrion
   B. Golgi apparatus
   C. rough and smooth endoplasmic reticulum
   D. lysosome
   E. plasma membrane

7. Which one of the following statements about prokaryotic and eukaryotic cells is *false*?
   A. Eukaryotic cells have membrane-enclosed organelles.
   B. Prokaryotic cells evolved before eukaryotic cells evolved.
   C. Eukaryotic cells are generally smaller than prokaryotic cells.
   D. Eukaryotic cells divide the labor of life among many internal compartments.
   E. Most bacteria are surrounded by a rigid cell wall.

8. Which one of the following has prokaryotic cells?
   A. a mushroom
   B. a bacterium
   C. an oak tree
   D. a crawfish
   E. a virus

9. True or False? Bacteria and archaea both consist of <u>prokaryotic</u> cells.

10. True or False? The genetic material of eukaryotic cells is housed in the <u>endoplasmic reticulum</u>.

11. Small membrane-bound parts of eukaryotic cells with specific functions are called ____________.

12. The magnification of a transmission electron microscope is about ____________ times greater than that of a typical light microscope.

13. Cell surfaces are best revealed by a(n) ____________ electron microscope.

14. Internal details of cells are best revealed by a(n) ____________ electron microscope.

15. A nucleus bordered by a membranous envelope is found in ____________ cells, but not in ____________ cells.

16. The generalization that all living things are composed of cells is called ____________ theory.

17. Some prokaryotic cells have ____________, which help attach them to surfaces.

18. In a eukaryotic cell, the region between the nucleus and the plasma membrane is the ____________.

19. One difference between plant and animal cells is that only plant cells have ____________ that convert light energy to the chemical energy of food.

Membrane Structure

20. The membranes of cells are primarily composed of:
    A. lipids and nucleic acids.
    B. proteins and nucleic acids.
    C. lipids and carbohydrates.
    D. proteins and carbohydrates.
    E. lipids and proteins.

21. Which one of the following statements about cellular membranes is *false*?
    A. Many membrane molecules can move freely past one another.
    B. Most membranes have carbohydrates embedded in the phospholipid bilayer.
    C. Diverse proteins float like icebergs in the phospholipid sea.
    D. The structure of the plasma membrane is similar to the structure of the other internal membranes of eukaryotic cells.
    E. Phospholipids in the membrane form a bilayer.

22. Which one of the following statements about plant cell walls is *false*? Plant cell walls:
    A. help the cells maintain their shape.
    B. protect the cells from physical damage.
    C. restrain the cells from absorbing too much water.
    D. are the primary site of photosynthesis.

23. MRSA infections:
    A. could be prevented by the use of the antibiotic methicilin.
    B. are caused by a newly mutated virus.
    C. are typically treated successfully with penicillin.
    D. use a protein called PSM to disable human immune cells.

24. The phospholipids and most of the proteins in a membrane are free to drift about in what is called the fluid ____________.

25. The surfaces of most animal cells contain cell ____________, structures that connect them to another cell.

The Nucleus and Ribosomes: Genetic Control of the Cell

26. Which one of the following statements about the nucleus and ribosomes of eukaryotic cells is *false*?
    A. Chromosomes are composed of long strands of DNA attached to certain proteins.
    B. Pores in the nuclear envelope allow messenger RNA to move from the nucleus to the cytoplasm.
    C. Ribosomes are constructed in the cytoplasm from parts produced in the nucleolus.
    D. DNA moves from the nucleus to the cytoplasm to direct the production of proteins.
    E. Ribosomes may work either suspended in the cytosol or attached to the endoplasmic reticulum.

27. If we think of the cell as a factory, then the nucleus is its executive boardroom and the top managers are the:
    A. genes.
    B. ribosomes.
    C. chromosomes.
    D. mitochondria.
    E. endoplasmic reticulum.

28. True or False? The directions to make a protein move from <u>DNA</u> to <u>messenger RNA</u> and then to <u>ribosomes</u>.

29. Proteins are made on ribosomes in the cytoplasm using the directions from a(n) ____________ molecule.

30. Differences in the structures of bacterial and eukaryotic ____________ allow humans to use antibiotics that do not harm human cells.

The Endomembrane System: Manufacturing and Distributing Cellular Products

31. Which one of the following statements about the components of the endomembrane system is *false*?
    A. Products from the endoplasmic reticulum are modified in the Golgi apparatus.
    B. Lysosomes have several types of digestive functions.
    C. Lipids are typically produced by the rough endoplasmic reticulum.
    D. Central vacuoles contribute to plant growth and may contain pigments and poisons.
    E. The smooth endoplasmic reticulum helps detoxify poisons.

32. Examine Figures 4.14–4.18 of your text. Which one of the following sequences best represents the steps an enzyme would take from initial production to joining with a food vacuole?

    A. rough endoplasmic reticulum, Golgi apparatus, lysosome
    B. Golgi apparatus, rough endoplasmic reticulum, lysosome
    C. lysosome, smooth endoplasmic reticulum, Golgi apparatus
    D. smooth endoplasmic reticulum, Golgi apparatus, lysosome

33. The smooth endoplasmic reticulum:

    A. synthesizes lipids and helps to detoxify drugs.
    B. synthesizes carbohydrates and proteins.
    C. is the primary site of ribosome production.
    D. regulates the import of drugs and hormones.
    E. helps to destroy lysosomes.

34. True or False? The final destination of some proteins in the cell is determined by chemical tags applied by the <u>rough endoplasmic reticulum</u>.

35. Found in animal cells but absent from most plant cells, ____________ are membrane-bounded sacs of digestive enzymes.

36. Membrane proteins and secretory proteins are the typical products of the ____________ endoplasmic reticulum.

Chloroplasts and Mitochondria: Energy Conversion

37. In a plant cell, where is light energy trapped and converted to chemical energy?

    A. the inner mitochondrial membrane
    B. the outer mitochondrial membrane
    C. the outer chloroplast membrane
    D. the stroma of a chloroplast
    E. the grana in the chloroplast

38. Compare Figures 4.19 and 4.20 of your text. Which of the following pairs of structures are the largest membranes in chloroplasts and mitochondria?

    A. outer chloroplast membrane and outer mitochondrial membrane
    B. inner chloroplast membrane and outer mitochondrial membrane
    C. outer chloroplast membrane and inner mitochondrial membrane
    D. the grana of chloroplasts and the outer mitochondrial membrane
    E. the grana of chloroplasts and the inner mitochondrial membrane

39. True or False? Plant cells have <u>chloroplasts and mitochondria</u>.

40. Cells use molecules of ____________ as the direct energy source for most of their work.

41. The extensive folds of the inner mitochondrial membrane are called ____________.

The Cytoskeleton: Cell Shape and Movement

42. Which one, if any, of the following is *not* a function of the cytoskeleton?
    - A. gives mechanical support to the cell
    - B. anchors organelles
    - C. helps a cell maintain its shape
    - D. helps a cell move
    - E. All of the above are functions of the cytoskeleton.

43. Which one of the following statements about the cilia and flagella is *false*?
    - A. Cilia are generally shorter than flagella.
    - B. Cilia are more numerous than flagella.
    - C. Cilia and flagella have the same basic architecture.
    - D. Flagella line the inside of your windpipe.
    - E. Human sperm rely on flagella for movement.

44. True or False? A specialized arrangement of <u>microfilaments</u> helps move cilia and flagella.

45. The movement of dividing chromosomes is guided by ____________.

Evolution Connection: The Evolution of Antibiotic Resistance

46. How do antibiotic-resistant bacteria evolve?
    - A. Antibiotics cause bacteria to change and become resistant.
    - B. The bacteria need to change to survive. Thus, bacteria evolve resistance.
    - C. Antibiotics cause mutations that lead to antibiotic resistance.
    - D. Natural selection favors individual bacteria with beneficial random mutations.

47. True or False? Antibiotics are not effective against <u>viruses</u>.

Word Roots

chloro = green (chloroplast: the green organelle of photosynthesis)
chromo = color (chromosome: a thread-like, darkly staining structure packaging DNA in the nucleus)
cili = small hair (cilium: a short, hairlike cellular appendage with a microtubule core)
cyto = cell (cytoplasm: cell region between the nucleus and the plasma membrane)
endo = inner (endomembrane system: an internal system of membranous organelles)
eu = true (eukaryotic: cell type with a membrane-enclosed nucleus and other organelles)
extra = outside (extracellular: the substance around animal cells)
flagell = whip (flagellum: a long, whiplike cellular appendage that moves cells)
micro = small (microtubules: microscopic tubular filaments contributing to the cytoskeleton)

plasm = molded (plasma membrane: the thin layer that sets a cell apart from its surroundings)

pro = before (prokaryotic: the first cells, lacking a membrane-enclosed nucleus and other organelles)

reticul = network (endoplasmic reticulum: membranous network where proteins are produced)

trans = across (transport vesicles: membranous spheres that move materials across a cell)

vacu = empty (vacuole: sac that buds from the ER, Golgi apparatus, or plasma membrane)

Key Terms

cell junctions
cell theory
central vacuole
chloroplast
chromatin
chromosome
cilia
cristae
cytoplasm
cytoskeleton
electron microscope (EM)
endomembrane system
endoplasmic reticulum (ER)
eukaryotic cell
extracellular matrix
flagella
fluid mosaic
food vacuoles
Golgi apparatus
grana
light microscope (LM)
lysosome
magnification
matrix
microtubule
mitochondria
nuclear envelope
nucleolus
nucleus
organelles
phospholipid
phospholipid bilayer
plasma membrane
prokaryotic cell
resolving power
ribosome
rough ER
scanning electron microscope (SEM)
smooth ER
stroma
transmission electron microscope (TEM)
transport vesicles
vacuole

Crossword Puzzle

Use the Key Terms list from this chapter to fill in the crossword puzzle.

ACROSS

1. a type of cell found only in the bacteria and archaea
4. a network of interconnected membranous sacs studded with ribosomes
5. a double membrane, perforated with pores, that encloses the nucleus
7. a measure of the clarity of an image
8. the thickest of the three main kinds of fibers making up a eukaryotic cytoskeleton
9. a membrane fat with a hydrophilic "head" and two hydrophobic "tails"
11. an atom's central core, containing protons and neutrons
13. an organelle that modifies, stores, and ships products of the endoplasmic reticulum
15. the place in a eukaryotic cell nucleus where ribosomes are made
16. an increase in the apparent size of an object
17. the thick fluid within the chloroplast
19. an organelle that functions as the site of protein synthesis in the cytoplasm
20. a digestive organelle in eukaryotic cells
22. a description of the flowing nature of membrane structure
23. membranous spheres that bud from the endoplasmic reticulum
25. a type of theory that all living things are composed of cells and all cells come from other cells
27. initials for an optical instrument with lenses that bend visible light to magnify images
28. cellular appendages similar to flagella, it moves some protists through water
29. a type of cell that has a membrane-enclosed nucleus and organelles
30. the combination of DNA and proteins that constitutes eukaryotic chromosomes
31. initials for an instrument that focuses an electron beam to see fine details
32. a membrane-enclosed sac occupying most of the interior of a mature plant cell
33. everything inside a cell between the plasma membrane and the nucleus
34. a photosynthetic organelle found in plants and some protists
35. initials for a type of microscope that uses a penetrating electron beam to study internal cellular details
36. infoldings of the inner mitochondrial membrane
38. a type of system of membranous organelles subdividing eukaryotic cells
39. a membrane-enclosed sac in a eukaryotic cell with diverse functions
40. a double layer that makes up the basic fabric of biological membranes

DOWN

1. the thin layer of lipids and proteins that sets a cell off from its surroundings
2. a structure with a specialized function within a cell
3. initials for a type of microscope that uses an electron beam to study a specimen's surface
6. a sticky coat secreted by most animal cells
10. the simplest type of digestive cavity, found in protists
12. a sticky coat outside of cells
14. interconnected stacks where sunlight is captured during photosynthesis
18. the organelle in eukaryotic cells where cellular respiration occurs
21. similar to cilia in structure, these cellular appendages propel protists and sperm
24. a type of connection between cells
26. a network of fine fibers that provides structural support for a eukaryotic cell
28. a threadlike, gene-carrying structure found in the nucleus of a eukaryotic cell

DOWN

37. a membranous network of tubules not associated with ribosomes in a eukaryotic cell
38. initials for a membranous organelle connected to the outer nuclear membrane

CHAPTER 5

The Working Cell

Studying Advice

a. This chapter addresses many abstract ideas that may initially be difficult to grasp. Do not try to understand all of this chapter in a single night. Read the material slowly and carefully and study the figures as you go along. Take frequent study breaks and review what you have already read before continuing further into the chapter.

b. Many of the concepts in this chapter relate to events in your life. It is always easier to remember a new idea by relating it to something you already know or have experienced. Look for these connections as you study.

Student Media

Activities

Energy Transformations
The Structure of ATP
How Enzymes Work
Membrane Structure
Diffusion
Facilitated Diffusion
Osmosis and Water Balance in Cells
Active Transport
Exocytosis and Endocytosis
Cell Signaling

BioFlix

Membrane Transport

Biology Labs On-Line

EnzymeLab

BLAST Animations

Structure of ATP
ATP/ADP Cycle
How Enzymes Work: Activation Energy
Enzymes Regulation: Competitive Inhibition
Diffusion
Passive Diffusion Across a Membrane
Active Transport: Sodium-Potassium Pump
Endocytosis and Exocytosis

LabBench

Enzyme Catalysis
Diffusion and Osmosis

MP3 Tutors

Basic Energy Concepts

Process of Science

How Is the Rate of Enzyme Catalysis Measured?
How Do Salt Concentrations Affect Cells?
How Do Cells Communicate with Each Other?
What Determines if Water Moves Into or Out of a Plant Cell?

Videos

Discovery Channel Video: Cells
Plasmolysis
Turgid *Elodea*

Organizing Tables

TABLE 5.1 Compare the reactants, products, and efficiencies of cellular respiration and the use of gasoline in an engine by completing this table.

	Reactants	Products	Efficiency (Percent of energy used to do work)
Cellular respiration			
Burning of gasoline in an automobile engine			

TABLE 5.2 Compare the processes of passive transport, facilitated diffusion, and active transport.

	Passive Transport	Facilitated Diffusion	Active Transport
Does the process require the use of ATP?			
What special membrane proteins, if any, are needed?			
Describe an example in a cell.			

TABLE 5.3 Compare the reactions of animal and plant cells when placed into isotonic, hypotonic, and hypertonic solutions. Compare your results to Figure 5.14 of your text.

	Isotonic Solution	Hypotonic Solution	Hypertonic Solution
Animal cell			
Plant cell			

Content Quiz

Directions: Identify the *one* best answer for the multiple-choice questions. For true/false questions, determine if the statement is true or false. If false, change the underlined word(s) to make the statement true. Finally, add the correct word(s) to the fill-in-the-blank questions to make the statements true.

Biology and Society: Natural Nanotechnology

1. Which one of the following is *not* one of the three main concerns in the lives of working cells?
   A. energy
   B. plasma membrane
   C. enzymes
   D. nitrogen

2. True or False? Researchers attached ATP from the process of glycolysis to a computer chip and found that the enzymes still functioned.

3. All living processes depend on ____________.

Some Basic Energy Concepts

CONSERVATION OF ENERGY

4. You are riding a bike up and down hills. At which point do you have the greatest potential energy?
   A. at the bottom of the hill
   B. at the top of the hill
   C. riding down the hill at the fastest speed
   D. climbing the hill

5. True or False? Energy is defined as the <u>capacity to do work</u>.

6. The principle known as ____________ states that it is not possible to destroy or create energy.

ENTROPY

7. All energy conversions:
   A. destroy some energy.
   B. decrease the entropy of the universe.
   C. generate some heat.
   D. produce ATP.

8. True or False? The heat produced by an automobile engine is a type of <u>kinetic energy</u>.

9. The term ____________ is used as a measure of disorder or randomness.

CHEMICAL ENERGY

10. We feel warmer when we exercise because of extra heat produced by:
    A. cellular respiration.
    B. breathing faster.
    C. friction of blood flowing through the body.
    D. our movement through the air around us.
    E. sweating.

11. Examine Figure 5.2 of your text. Which one of the following is *not* produced by the engine and cellular respiration?
    A. heat
    B. water
    C. oxygen
    D. carbon dioxide

12. True or False? The chemical energy in molecules of glucose and other fuels is a special type of <u>kinetic</u> energy.

13. Molecules of carbohydrates, fats, and gasoline all have structures that make them especially rich in ____________ energy.

FOOD CALORIES

14. The energy in a 300-Calorie cheeseburger could raise the temperature of:
    A. 1 kilogram (kg) of water by 30°C.
    B. 3 kg of water by 100°C.
    C. 3 kg of water by 1°C.
    D. 30 kg of water by 30°C.
    E. 30 kg of water by 100°C.

15. True or False? The calories in food are a form of <u>potential</u> energy.

16. The amount of energy needed to raise 1 gram of water 1°C is a(n) ____________ while 1,000 times this amount of energy is a(n) ____________.

ATP and Cellular Work

THE STRUCTURE OF ATP, PHOSPHATE TRANSFER

17. Moving ions across cell membranes is an example of what type of ATP work?
    A. chemical work
    B. transport work
    C. neutral work
    D. mechanical work

18. True or False? ATP has <u>less</u> potential energy than ADP.

19. The energy for most cellular work is found in the ____________ of the ATP molecule.

THE ATP CYCLE

20. What happens to the ADP molecule produced when ATP loses a phosphate during an energy transfer?
    A. ADP is used to build carbohydrates, fats, and proteins for use throughout the cell.
    B. ADP is released from the cells.
    C. ADP is broken down further into carbon atoms that are then reassembled into proteins.
    D. Energy from cellular respiration is used to convert ADP back to ATP.
    E. None of the above statements are correct.

21. True or False? The energy used to regenerate a cell's supply of ATP comes from <u>cellular respiration</u>.

22. To generate ATP, energy is used to join a phosphate group with ____________.

Enzymes

ACTIVATION ENERGY

23. Which one of the following can be used to initiate a chemical reaction by lowering the activation energy?
    A. ATP
    B. oxygen
    C. phosphate groups
    D. carbon dioxide
    E. enzymes

24. True or False? Enzymes <u>raise</u> the activation energy of a chemical reaction.

25. The many chemical reactions that occur in organisms are collectively called ____________.

THE PROCESS OF SCIENCE: CAN ENZYMES BE ENGINEERED?

26. Researchers demonstrated that they could cause mutations in the ____________ that produced lactase to produce a new enzyme.
    A. gene
    B. protein
    C. enzyme
    D. cell

27. True or False? Researchers were able to use <u>natural</u> selection to produce a new type of enzyme.

28. The process of deliberately mutating a gene to produce a new enzyme is called ____________ evolution.

INDUCED FIT

29. When a substrate molecule slips into an active site, the active site changes shape slightly to embrace the substrate and catalyze the reaction. This interaction is called:
    A. substrate locking.
    B. active engagement.
    C. active site shifting.
    D. induced fit.
    E. substrate embracing.

30. True or False? Most enzymes are named for their <u>inhibitors</u>.

31. The reactant molecule called the ____________ binds at an enzyme's ____________.

ENZYME INHIBITORS

32. Examine the inhibitor in Figure 5.8b of your text. The inhibitor functions most like a:
    A. teacher giving you instructions for a laboratory exercise.
    B. friend who helps you do your laundry.
    C. person who has parked in your parking spot.
    D. screwdriver used to tighten screws.
    E. traffic light indicating to a driver when it is safe to cross an intersection.

33. True or False? The inhibition of an enzyme by its product is an example of <u>positive reinforcement</u>.

34. Many antibiotics that kill disease-causing bacteria are also enzyme ____________.

Membrane Function

PASSIVE TRANSPORT: DIFFUSION ACROSS MEMBRANES

35. Releasing fish into a pond and seeing them swim in all directions is most like the process of:
    A. diffusion.
    B. osmosis.
    C. osmoregulation.
    D. active transport.
    E. pinocytosis.

36. Which one of the following most relies upon the process of diffusion?
    A. recognizing a friend eating at another table in a cafeteria
    B. discussing a subject in preparation for an exam
    C. checking your pulse after exercise
    D. listening to a song on the radio
    E. selecting a new perfume or cologne

37. True or False? Cells <u>do not</u> need to spend energy for diffusion to occur.

38. A cell does not have to use ____________ for passive transport to occur.

39. Some substances have properties that prevent them from directly diffusing across a membrane. Such substances, however, can diffuse across a membrane with the help of specific transport proteins in the process of ____________.

OSMOSIS AND WATER BALANCE

40. If you soak your hands in dishwater, you may notice that your skin soaks up water and swells into distinct wrinkles. This is because your skin cells are ____________ to the ____________ dishwater.
    A. hypotonic; hypertonic
    B. hypertonic; hypotonic
    C. hypotonic; hypotonic
    D. isotonic; hypotonic
    E. hypertonic; isotonic

41. You decide to buy a new angelfish for your freshwater aquarium. When you introduce the fish into its new tank, the fish swells up and dies. You later learn it was an angelfish from the ocean. The unfortunate fish went from a ____________ solution in the ocean to a(n) ____________ solution in your freshwater aquarium.
    A. mesotonic; hypotonic
    B. hypertonic; isotonic
    C. hypertonic; hypotonic
    D. hypotonic; isotonic

42. Examine Figure 5.14 of your text. Why does the animal cell but not the plant cell burst when it is in a hypotonic solution?
    A. The plant cell is less hypertonic.
    B. The plant cell does not absorb as much salt.
    C. The cell wall keeps the plant cell from bursting.
    D. The water cannot easily move through the plant cell wall.

43. True or False? A crab in the ocean has the same salt concentration in its body as the surrounding seawater. Thus, the ocean is <u>hypotonic</u> to the crab.

44. When a *Paramecium* uses its contractile vacuole to remove excess water, it is engaged in the process called ____________.

ACTIVE TRANSPORT: THE PUMPING OF MOLECULES ACROSS MEMBRANES

45. Moving a molecule across a membrane and against its concentration gradient requires:
    A. phospholipids using passive transport.
    B. phospholipids using active transport.
    C. membrane transport proteins using passive transport.
    D. membrane transport proteins using active transport.
    E. membrane transport proteins using receptor-mediated endocytosis.

46. True or False? Membrane proteins using <u>passive</u> transport require ATP as a source of energy.

47. A nerve cell maintains a higher concentration of potassium ions inside itself than in its surroundings. This nerve cell is most likely using ____________ transport to keep the potassium levels high.

EXOCYTOSIS AND ENDOCYTOSIS: TRAFFIC OF LARGE MOLECULES

48. Which one of the following is a type of endocytosis in which very specific molecules are brought into a cell using specific membrane protein receptors?
   A. osmoregulation
   B. phagocytosis
   C. pinocytosis
   D. receptor-mediated endocytosis
   E. None of the above.

49. True or False? Tears and saliva are products released by cells using <u>endocytosis</u>.

50. In ____________, the cell gulps droplets of fluid by forming tiny vesicles.

THE ROLE OF MEMBRANES IN CELL SIGNALING

51. In a cell, the ____________ relays a signal and converts it to chemical forms that work within the cell.
   A. process of exocytosis
   B. process of endocytosis
   C. process of pinocytosis
   D. receptor protein
   E. signal transduction pathway

52. True or False? The three stages of cell signaling are <u>reception</u>, <u>transduction</u>, and <u>response</u>.

53. In a signal transduction pathway, external signals are received by specific receptor ____________.

Evolution Connection: The Origin of Membranes

54. Which one of the following is a component of all living cells?
   A. plasma membranes
   B. chloroplasts
   C. cell walls
   D. vacuoles

55. True or False? The tendency of lipids in water to spontaneously form membranes has led biomedical engineers to produce artificial vesicles called <u>liposomes</u>.

56. The key ingredient of membranes, ____________ spontaneously assemble into bilayers.

Word Roots

endo = within, inner (endocytosis: taking material into a cell)
exo = outside (exocytosis: eliminating some materials outside of a cell)
hyper = excessive (hypertonic: in comparing two solutions, it refers to the one with the greater concentration of solutes)
hypo = lower (hypotonic: in comparing two solutions, it refers to the one with the lower concentration of solutes)
iso = same (isotonic: solutions with equal concentrations of solutes)
kinet = move (kinetic: type of energy, it is the energy of motion)
phago = eat (phagocytosis: cellular eating)
pino = drink (pinocytosis: cellular drinking)
tonus = tension (isotonic: solutions with equal concentrations of solutes)

Key Terms

activation energy
active site
active transport
ADP
ATP
calorie
chemical energy
concentration gradient
conservation of energy
diffusion
endocytosis
energy
entropy
enzyme
enzyme inhibitor
exocytosis
facilitated diffusion
feedback regulation
heat
hypertonic
hypotonic
induced fit
isotonic
kinetic energy
metabolism
osmoregulation
osmosis
passive transport
phagocytosis
pinocytosis
plasmolysis
potential energy
receptor-mediated endocytosis
signal-transduction pathway
substrate
transport proteins

Crossword Puzzle

Use the Key Terms list from this chapter to fill in the crossword puzzle.

ACROSS

1. diffusion across a membrane without the input of energy
3. a type of pathway that converts a signal on a cell's surface into an inner response
6. protein that serves as a biological catalyst
8. the principle that energy can neither be created nor destroyed
11. a molecule composed of adenosine and three phosphate groups
12. the type of gradient in which a substance's density changes
15. the passive transport of water across a selectively permeable membrane
17. the type of energy stored in the chemical bonds of molecules
18. cellular eating
20. the type of energy that is stored
21. a measure of disorder, or randomness
23. the interaction between a substrate molecule and the active site of an enzyme
24. of two solutions, the one with the greater concentration of solutes
25. chemical that interferes with an enzyme's activities
26. the amount of energy associated with the movement of the atoms and molecules in a body of matter
27. the energy of motion
28. the type of protein that helps move substances across a cell membrane
29. of two solutions, the one with the lesser concentration of solutes
30. a specific substance (reactant) on which an enzyme acts
31. the passage of a substance across a biological membrane down its concentration gradient

DOWN

2. the control of water and solute balance in an organism
4. the type of endocytosis in which specific molecules move into a cell by inward budding
5. the many chemical reactions that occur in organisms
7. the capacity to perform work
8. the amount of energy that raises the temperature of 1 gram of water by 1 degree C
9. a molecule composed of adenosine and two phosphate groups
10. the movement of materials into the cytoplasm of a cell via membranous vesicles
13. a metabolic control in which the product inhibits the process that produced it
14. plant cell shriveling when there is a shortage of water
16. the tendency of molecules to move from high to low concentrations
19. having the same solute concentration as another solution
21. the movement of materials out of the cytoplasm of a cell via membranous vesicles
22. cellular drinking

CHAPTER 6

Cellular Respiration: Obtaining Energy from Food

Studying Advice

a. This chapter addresses many abstract ideas that may initially be difficult to grasp. Do not try to understand all of this chapter in a single night. Read the material slowly and carefully and study the figures as you go along. Take frequent study breaks and review what you have already read before continuing further into the chapter.

b. The following organizing tables should help you understand the basics of cellular respiration. Focus first on the overall process. Then read to gather the details. Table 6.1 should help you organize the definitions of the many pairs of contrasting terms.

Student Media

Activities

Build a Chemical Cycling System
Overview of Cellular Respiration
Glycolysis
The Citric Acid Cycle
Electron Transport
Fermentation

BioFlix

Cellular Respiration

Biology Labs On-Line

MitochondriaLab

BLAST Animations

Harvesting Energy: Krebs Cycle

LabBench

Cell Respiration

MP3 Tutors

Cellular Respiration Part 1—Glycolysis
Cellular Respiration Part 2—Citric Acid Cycle and Electron Transport

Process of Science

How Is the Rate of Cellular Respiration Measured?

Videos

Discovery Channel Video: Space Plants
Discovery Channel Video: Tasty Bacteria

Organizing Tables

TABLE 6.1 Compare the definitions of the pairs of terms.

Photosynthesis vs. ____________________ Respiration
Autotrophs vs. ____________________ Heterotrophs
Producers vs. ____________________ Consumers
Aerobic vs. ____________________ Anaerobic

TABLE 6.2 **Compare the processes of glycolysis, the citric acid cycle, and electron transport in this table. Refer to the "Input" and "Output" portions of Figures 6.7, 6.9, and 6.10 and the ATP yields in Figure 6.13.**

	Location	Reactants	Products	Energy Yield (Number of ATP produced)
Glycolysis				
Citric acid cycle				
Electron transport chain				

TABLE 6.3 **Compare the products of each of the following reactions.**

Process	Products
Fermentation in human muscle cells	
Alcoholic fermentation in yeast	

Content Quiz

Directions: Identify the *one* best answer for the multiple-choice questions. For true/false questions, determine if the statement is true or false. If false, change the underlined word(s) to make the statement true. Finally, add the correct word(s) to the fill-in-the-blank questions to make the statements true.

Biology and Society: Marathoners versus Sprinters

1. Compared to fast-twitch muscle fibers, slow-twitch fibers:
   A. fatigue more quickly.
   B. use more oxygen to make ATP.
   C. do not require a blood supply.
   D. only contract when we are sleeping.
   E. are only found in muscles of the legs.

2. True or False? All human muscles contain fast-twitch and slow-twitch fibers.

3. Sprinters have muscles that contain higher percentages of ____________ fibers.

Energy Flow and Chemical Cycling in the Biosphere

PRODUCERS AND CONSUMERS

4. A deer chewing on grass is an example of a ____________ eating a ____________.

   A. producer; consumer
   B. consumer; producer
   C. producer; heterotroph
   D. consumer; heterotroph

5. True or False? Photosynthesis uses light energy from the sun to power chemical processes that build <u>inorganic</u> molecules.

6. True or False? Plants and other photosynthetic organisms are the <u>consumers</u> in an ecosystem.

7. Organisms that *cannot* make organic molecules from inorganic ones are called ____________.

CHEMICAL CYCLING BETWEEN PHOTOSYNTHESIS AND CELLULAR RESPIRATION

8. Plants use photosynthesis to join:
   A. carbon dioxide and water to make oxygen and glucose.
   B. carbon dioxide and glucose to make water and oxygen.
   C. carbon dioxide and oxygen to make water and glucose.
   D. glucose and oxygen to make water and carbon dioxide.
   E. water and oxygen to make glucose and carbon dioxide.

9. Cellular respiration:
   A. occurs in chloroplasts.
   B. is used by plants, but not animals.
   C. is part of photosynthesis.
   D. harvests energy stored in sugars and other organic molecules.
   E. produces glucose and oxygen.

10. Examine Figure 6.2 of your text. In this ecosystem, the carbon dioxide that the plants use in photosynthesis came from:
    A. other plants.
    B. a fox.
    C. a rabbit.
    D. All of the above.
    E. None of the above.

11. True or False? Cellular respiration occurs in plant and animal cells in organelles called <u>chloroplasts</u>.

12. Cellular respiration typically uses energy extracted from organic fuel to produce another source of chemical energy called ____________.

13. Photosynthesis combines together ____________ and ____________, which are products of cellular respiration.

Cellular Respiration: Aerobic Harvest of Food Energy

14. What do cellular respiration and breathing have in common? Both processes:
    A. produce ATP.
    B. produce glucose.
    C. take in carbon dioxide and release oxygen.
    D. take in oxygen and release carbon dioxide.

15. True or False? An aerobic cellular process requires <u>carbon dioxide</u>.

16. The aerobic harvesting of chemical energy from organic food molecules defines ____________.

THE OVERALL EQUATION FOR CELLULAR RESPIRATION

17. Which one of the following is *not* produced because of the breakdown of a single glucose molecule by cellular respiration?
    A. 6 molecules of carbon dioxide
    B. 6 molecules of water
    C. 6 molecules of oxygen
    D. up to 38 ATP molecules

18. True or False? Cellular respiration consists of <u>many</u> steps.

19. Oxygen is a vital part of aerobic respiration because it accepts ____________ atoms from glucose.

THE ROLE OF OXYGEN IN CELLULAR RESPIRATION

20. Which one of the following occurs during the transfer of hydrogen in cellular respiration?
    A. Glucose is reduced, oxygen is oxidized, and energy is released.
    B. Glucose is oxidized, oxygen is reduced, and energy is released.
    C. Glucose is reduced, oxygen is oxidized, and energy is consumed.
    D. Glucose is oxidized, oxygen is reduced, and energy is consumed.

21. Which one of the following does *not* occur during the cellular respiration of glucose?

    A. NAD^+ donates electrons to NADH.

    B. NADH transfers electrons from glucose to the top of the electron transport chain.

    C. Electrons cascade down the electron transport chain, giving up a small amount of energy with each transfer.

    D. Oxygen is the final electron acceptor at the bottom of the electron transport chain.

22. Examine Figure 6.5 of your text. The transfer of electrons during cellular respiration is most like:

    A. a bowling ball striking a set of pins.

    B. walking down a set of stairs.

    C. climbing a mountain.

    D. running up a set of stairs.

    E. playing catch with a baseball.

23. True or False? In the electron transport chain, <u>NAD^+</u> functions like gravity, pulling electrons down the chain.

24. NADH undergoes the process of ____________ when it donates its electrons to the electron transport chain.

AN OVERVIEW OF CELLULAR RESPIRATION

25. Which one of the following is the correct sequence of events in cellular respiration?

    A. glycolysis, electron transport chain, citric acid cycle

    B. glycolysis, citric acid cycle, electron transport chain

    C. citric acid cycle, glycolysis, electron transport chain

    D. electron transport chain, glycolysis, citric acid cycle

    E. electron transport chain, citric acid cycle, glycolysis

26. True or False? Cellular respiration involves <u>more than two dozen</u> reactions.

27. The inner mitochondrial membrane is the site of the ____________.

THE THREE STAGES OF CELLULAR RESPIRATION

28. Which one of the following is most like water moving past a dam to generate electricity?

    A. glycolysis

    B. electron transport chain

    C. citric acid cycle

    D. None of the above.

29. Which one of the following statements about cellular respiration is *false*?
   A. ATP synthase, located within the inner mitochondrial membrane, helps form ATP.
   B. Cellular respiration is a metabolic pathway consisting of more than two dozen chemical reactions, each catalyzed by a specific enzyme.
   C. Inside mitochondria, glycolysis joins a pair of three-carbon pyruvic acid molecules to form a molecule of glucose.
   D. The citric acid cycle breaks down molecules of acetyl-CoA to release CO_2 and energy trapped by NADH.

30. The extensive infolding of the inner mitochondrial membrane is likely an adaptation to:
   A. make it more difficult for oxygen to pass through.
   B. increase the flexibility of mitochondria.
   C. make it more difficult for glucose to pass through.
   D. increase the surface area of the electron transport system.
   E. increase the surface area for glycolysis.

31. Examine Figure 6.10 of your text. What happens to the two carbons in acetyl CoA during the citric acid cycle? The two carbons are:
   A. attached to NADH.
   B. used to make glucose.
   C. released as CO_2.
   D. added to ADP to make ATP.
   E. destroyed in the process.

32. True or False? Most ATP is generated during <u>the process of glycolysis</u>.

33. In the citric acid cycle, coenzyme A joins ____________ to form acetyl CoA.

THE VERSATILITY OF CELLULAR RESPIRATION

34. Which of the following can be used to generate ATP by cellular respiration?
   A. glycerol
   B. fatty acids
   C. amino acids
   D. sugars
   E. All of the above.

35. True or False? Glycolysis and the citric acid cycle each contribute <u>two ATP</u> by direct synthesis.

36. One molecule of glucose can be broken down by cellular respiration to generate up to ____________ molecules of ATP.

Fermentation: Anaerobic Harvest of Food Energy

FERMENTATION IN HUMAN MUSCLE CELLS

37. Which one of the following statements about fermentation is *false*?
    A. Your cells can produce ATP when no oxygen is present.
    B. When your muscles use fermentation, glucose is produced.
    C. Fermentation in your cells uses the process of glycolysis.
    D. Fermentation produces two ATP per glucose molecule.
    E. Fermentation occurs when oxygen levels are not sufficient to support cellular respiration.

38. True or False? Fermentation is an <u>anaerobic</u> process.

39. During the process of fermentation in muscle cells, pyruvic acid is converted to ____________.

THE PROCESS OF SCIENCE: DOES LACTIC ACID BUILDUP CAUSE MUSCLE BURN?, FERMENTATION IN MICROORGANISMS

40. Yeast used to make beer and bread use fermentation to produce:
    A. ATP, carbon dioxide, and oxygen.
    B. ATP, water, and oxygen.
    C. ATP, ethyl alcohol, and carbon dioxide.
    D. ethyl alcohol and oxygen.
    E. carbon dioxide and water.

41. The buildup of lactic acid in muscle cells after hard exercise:
    A. causes muscle fatigue.
    B. adds another source of ATP to the cells.
    C. adds NADH to the system.
    D. has unclear effects.

42. True or False? Soy sauce, olives, pickles, sour cream, and yogurt are produced using <u>lactic acid</u> formed during fermentation.

43. Yeast is a microscopic ____________ capable of cellular respiration and ____________.

Evolution Connection: Life before and after Oxygen

44. Glycolysis is considered an ancient metabolic process because:
    A. oxygen was abundant in the early Earth atmosphere.
    B. glycolysis is found only in animals.
    C. glycolysis occurs in the cytosol.
    D. it produces more ATP per glucose molecule than cellular respiration.

45. True or False? The process of glycolysis may be <u>more than</u> 3 billion years old.

46. The early Earth atmosphere had very little ____________, favoring anaerobic organisms that could use glycolysis.

Word Roots

aero = air (aerobic: chemical reaction using oxygen)
auto = self; **troph** = food (autotroph: organism that makes its own organic matter)
glyco = sweet; **lysis** = split (glycolysis: process that splits glucose into two molecules)
hetero = other (heterotroph: Greek word that means "other feeder")
photo = light (photosynthesis: process using light energy to make organic molecules)

Key Terms

aerobic
anaerobic
ATP synthase
autotroph
cellular respiration
citric acid cycle
consumer
electron transport
electron transport chain
fermentation
glycolysis
heterotroph
NADH
oxidation
photosynthesis
producer
redox reaction
reduction

Crossword Puzzle

Use the Key Terms list from this chapter to fill in the crossword puzzle.

ACROSS

5. process using light energy to make organic molecules
8. abbreviation for type of reaction that transfers electrons
10. cluster of proteins built into the inner mitochondrial membrane
13. process that harvests energy stored in sugars
14. chemical reaction using oxygen
15. loss of electrons during a redox reaction
16. a type of chain composed of electron-carrier molecules used to make ATP

DOWN

1. type of cycle that continues breakdown of glucose after glycolysis
2. anaerobic harvest of food energy
3. organism that cannot make its own organic food molecules
4. type of organism that uses photosynthesis
6. a molecule that carries electrons from glucose to the electron transport chain
7. process that splits glucose into two molecules
9. acceptance of electrons during a redox reaction
10. organism that makes all of its own organic matter
11. type of chemical reaction that does not use oxygen
12. heterotroph that eats other heterotrophs or autotrophs

CHAPTER 7

Photosynthesis: Using Light to Make Food

Studying Advice

a. Like Chapters 5 and 6, this chapter addresses many abstract ideas that may initially be difficult to grasp. Do not try to understand this entire chapter in a single night. Read the material slowly and carefully and study the figures as you go along. Figures 7.9–7.13 are especially helpful in understanding the molecular details of photosynthesis. Take frequent study breaks and review what you have already read before continuing further into the chapter.

b. The following organizing tables should help you understand the two basic steps of photosynthesis and the different photosynthetic strategies used by plants.

Student Media

Activities

The Plants in Our Lives
The Sites of Photosynthesis
Overview of Photosynthesis
Light Energy and Pigments
The Light Reactions
The Calvin Cycle
Photosynthesis in Dry Climates

BioFlix

Photosynthesis

Biology Labs On-Line

LeafLab

BLAST Animations

Photosynthesis: Light-Independent Reactions

LabBench

Photosynthesis

MP3 Tutors

Photosynthesis

Process of Science

How Does Paper Chromatography Separate Plant Pigments?
How Is the Rate of Photosynthesis Measured?

Videos

Discovery Channel Video: Trees
Discovery Channel Video: Space Plants

Organizing Tables

TABLE 7.1 Compare the reactants, products, and location in the cell for the three reactions listed at the left in the table.

Reactants	Products	Location in the Cell
Cellular respiration		
Light reaction		
Calvin cycle		

TABLE 7.2 Compare the three strategies used by plants to incorporate carbon dioxide into the Calvin cycle.

	Source of Carbon Dioxide for the Calvin Cycle	Common Examples
C_3 plants		
C_4 plants		
CAM plants		

Content Quiz

Directions: Identify the *one* best answer for the multiple-choice questions. For true/false questions, determine if the statement is true or false. If false, change the underlined word(s) to make the statement true. Finally, add the correct word(s) to the fill-in-the-blank questions to make the statements true.

Biology and Society: Green Energy

1. Which of the following, if any, is *not* an advantage of using willows as biofuels? Willows:
   A. are fast-growing.
   B. do not need to be replanted after cutting.
   C. have very little of the sulfur compounds present in fossil fuels.
   D. provide habitat for wildlife and reduce erosion.
   E. All of the above *are* advantages of using willows as biofuels.

2. True or False? Biofuels are a renewable energy source.

3. True or False? As societies became industrialized, coal was largely displaced as an energy source.

4. The energy released from biofuels ultimately comes from ____________.

The Basics of Photosynthesis

CHLOROPLASTS: SITES OF PHOTOSYNTHESIS

5. Photosynthesis commonly occurs when sunlight hits chlorophyll pigments in the:
   A. thylakoid membrane of a mitochondrion in a leaf.
   B. stroma of a chloroplast in a cell of a leaf.
   C. grana of the nucleus of a cell in a leaf.
   D. thylakoid membrane of a chloroplast in a cell of a leaf.

6. Examine Figure 7.2 to compare the drawing and photograph of a chloroplast. How is the drawing different from the photograph of an actual chloroplast from a plant leaf?
   A. The grana are not stacked in the drawing.
   B. The grana are not interconnected in the drawing.
   C. There is no stroma in the drawing.
   D. The intermembrane space is much larger in the drawing.

7. True or False? The stacking of thylakoid membranes into grana decreases the surface area for photosynthesis.

8. Carbon dioxide enters the leaf, and oxygen exits, through tiny pores called ____________.

THE OVERALL EQUATION FOR PHOTOSYNTHESIS

9. Which one of the following statements about photosynthesis is *false*? Photosynthesis:
   - A. uses the products of respiration.
   - B. produces carbon dioxide and oxygen.
   - C. uses energy from sunlight to split water and release oxygen into the atmosphere.
   - D. boosts the energy in electrons "uphill" to join electrons to carbon dioxide.

10. True or False? Inside chloroplasts, carbon dioxide is split to form hydrogen and oxygen.

11. The sugar molecule produced by photosynthesis is ____________.

A PHOTOSYNTHESIS ROAD MAP

12. ATP and NADPH, produced by the:
    - A. light reaction, are used to reduce carbon dioxide to form glucose.
    - B. light reaction, are used to reduce glucose to form carbon dioxide.
    - C. Calvin cycle, are used to reduce carbon dioxide to form glucose.
    - D. Calvin cycle, are used to reduce glucose to form carbon dioxide.

13. Examine the road map for photosynthesis in Figure 7.3. Which one of the following is a product of the Calvin cycle?
    - A. ATP
    - B. NADPH
    - C. glucose
    - D. carbon dioxide
    - E. water

14. True or False? The light reaction does not produce sugar.

15. The Calvin cycle depends on a supply of ____________ and ____________ from the light reactions.

The Light Reactions: Converting Solar Energy to Chemical Energy

THE NATURE OF SUNLIGHT

16. Which one of the following statements about radiation is *false*?
    - A. Waves in the electromagnetic spectrum carry energy.
    - B. The electromagnetic spectrum is the full range of radiation.
    - C. A wavelength is the distance between the crests of two adjacent waves.
    - D. The wavelengths of visible light that we see were absorbed by what we are looking at.

17. True or False? We can see only a small portion of the entire electromagnetic spectrum.

18. A dark blue shirt will absorb ____________ light energy than a very light blue shirt.

THE PROCESS OF SCIENCE: WHAT COLORS OF LIGHT DRIVE PHOTOSYNTHESIS?

19. What did Engelmann use to determine where the highest concentrations of oxygen were produced?
    A. oxygen-sensitive bacteria
    B. an electronic oxygen probe
    C. signs of bubbles of oxygen gas accumulating
    D. the speed of particles moving away from the oxygen-producing algae
    E. a change in the color of the water around the oxygen-producing algae

20. True or False? Chloroplasts use <u>all</u> of the wavelengths of light to drive photosynthesis.

21. The wavelengths of light least used by plants form the color ____________.

CHLOROPLAST PIGMENTS

22. Chlorophyll *b*:
    A. absorbs mainly blue-violet and red light.
    B. participates directly in the light reactions.
    C. absorbs and dissipates excessive light that could damage chlorophyll.
    D. participates indirectly in the light reactions by transferring light energy to chlorophyll *a*.

23. True or False? Chloroplast pigments are all located within the <u>outer</u> membrane of the chloroplast.

24. Only the pigment ____________ participates directly in the light reactions.

HOW PHOTOSYSTEMS HARVEST LIGHT ENERGY

25. The photosynthetic pigments work together to function most like:
    A. a telephone.
    B. a brain.
    C. a mirror.
    D. an umbrella.
    E. an antenna.

26. Which one of the following is the general sequence of energy transfer during photosynthesis? Light energy in a photon is transferred first to:
    A. a primary electron acceptor, then to chlorophyll *a*, and finally to pigment molecules.
    B. pigment molecules, then to a primary electron acceptor, and finally to chlorophyll *a*.
    C. pigment molecules, then to chlorophyll *b*, and finally to a primary electron acceptor.
    D. pigment molecules, then to chlorophyll *a*, and finally to a primary electron acceptor.
    E. None of the above.

27. True or False? The reaction center of a photosystem consists of a chlorophyll *a* molecule next to a <u>primary electron acceptor</u>.

28. A fixed quantity of light energy is a(n) ____________.

HOW THE LIGHT REACTIONS GENERATE ATP AND NADPH

29. Which one of the following parts of a mitochondrion functions most like the thylakoid membrane?
    A. mitochondrial matrix
    B. outer mitochondrial membrane
    C. inner mitochondrial membrane
    D. the intermembrane space

30. The extensive infolding of the inner thylakoid membrane is likely an adaptation to:
    A. make it more difficult for oxygen to pass through.
    B. increase the flexibility of chloroplasts.
    C. make it more difficult for hydrogen ions to pass through.
    D. increase the surface area of the light reactions.
    E. increase the surface area for glycolysis.

31. Oxygen is produced during the:
    A. water-splitting photosystem.
    B. electron transport chain.
    C. NADPH-producing photosystem.
    D. Calvin cycle.

32. True or False? Both cellular respiration and photosynthesis have <u>ATP synthases</u> that use the energy stored by an H^+ gradient.

33. The electron transport chain in the chloroplast releases energy used to produce ____________.

The Calvin Cycle: Making Sugar from Carbon Dioxide

34. Which of the following are produced by the Calvin cycle?
    A. $NADP^+$, ADP + P, and G3P
    B. NADPH, ATP, and glucose
    C. NADPH, ADP + P, and glucose
    D. NADPH, ATP, and G3P
    E. glucose and oxygen

35. True or False? Plant cells use <u>G3P</u> to make glucose and other organic molecules they need.

36. The Calvin cycle functions within the ____________ of a chloroplast.

Evolution Connection: Solar-Driven Evolution

37. Which one of the following types of plants is most likely to be found in a desert?
    A. CAM plants
    B. C_4 plants
    C. wheat
    D. C_3 plants
    E. rice

38. True or False? The same adaptation that reduces water loss also prevents carbon dioxide from entering the <u>roots</u>.

39. CAM plants are most likely to have their ____________ open mainly at night.

40. During a hot and dry summer, C_3 plants can lose dangerous amounts of ____________ if their stomata remain open during the day.

Word Roots

chloro = green; **plast** = formed or molded (chloroplast: the organelle of photosynthesis)
electro = electricity; **magne** = magnetic (electromagnetic spectrum: the full range of radiation)
photo = light; **syn** = with or together (photosynthesis: the process whereby plants transform light energy into chemical energy)
phyll = leaf (chlorophyll: photosynthetic pigment in chloroplasts)
stoma = mouth (stomata: tiny pores in leaves through which gases are exchanged)
thylac = a sac or pouch (thylakoids: membranous sacs suspended in the stroma)

Key Terms

C_3 plants
C_4 plants
Calvin cycle
CAM plants
chlorophyll
chlorophyll *a*
chloroplast
electromagnetic spectrum
grana
light reactions
NADPH
photon
photosynthesis
photosystem
primary electron acceptor
reaction center
stomata
stroma
thylakoids
wavelength

Crossword Puzzle

Use the Key Terms list from this chapter to fill in the crossword puzzle.

ACROSS

2. the type of reaction converting solar energy to chemical energy
3. a fixed quantity of light energy
6. membranous sacs suspended in the stroma
7. pigment that absorbs blue-violet and red light
9. the process whereby plants transform light energy into chemical energy
10. stacks of hollow disks inside chloroplasts
11. tiny pores in leaves through which gases are exchanged
12. the type of plant that uses crassulacean acid metabolism
13. a light-absorbing pigment in the chloroplasts
14. a type of electron acceptor that harvests light
15. the type of spectrum of all types of radiation
18. distance between crests of adjacent light waves
19. thick fluid deep inside a chloroplast

DOWN

1. the organelle of photosynthesis
4. cluster of pigment molecules
5. electron carrier involved in photosynthesis
7. process that makes sugar from carbon dioxide
8. in a chloroplast, the chlorophyll *a* molecule and primary electron acceptor
16. type of plant that uses the Calvin cycle to form three-carbon compounds as the first stable intermediate
17. type of plant that prefaces the Calvin cycle with reactions that incorporate carbon dioxide into four-carbon compounds

CHAPTER 8

Cellular Reproduction: Cells from Cells

Studying Advice

a. Read carefully and study Figures 8.7 and 8.14 of your text describing the key events of mitosis and meiosis. Figure 8.15 is a very useful comparison of both processes.

b. The key to understanding mitosis and meiosis is learning how the chromosomes are arranged and how they separate during metaphase. Again, Figures 8.7 and 8.14 of your text are crucial.

c. The following organizing tables should help you sort out the important details of the cell cycle, mitosis, and meiosis.

Student Media

Activities

Asexual and Sexual Life Cycles
The Cell Cycle
Mitosis and Cytokinesis Animation
Mitosis and Cytokinesis Video
Causes of Cancer
Human Life Cycle
Meiosis Animation
Origins of Genetic Variation

BioFlix

Mitosis
Meiosis

BLAST Animations

Cytokinesis in Plant Cells
Genetic Variation: Independent Assortment
Genetic Variation: Fusion of Gametes

LabBench

Mitosis and Meiosis

MP3 Tutors

Mitosis
Meiosis
Comparing Mitosis and Meiosis

Process of Science

How Much Time Do Cells Spend in Each Phase of Mitosis?
How Can the Frequency of Crossing Over Be Estimated?

Videos

Discovery Channel Video: Cells
Discovery Channel Video: Fighting Cancer
Hydra Budding
Animal Mitosis (time-lapse)
Sea Urchin Embryonic Development (time-lapse)

Organizing Tables

TABLE 8.1 Compare the key events of each stage of the cell cycle. Figure 8.6 will help.

	Key Events
M phase	
G_1 phase	
G_2 phase	
S phase	

TABLE 8.2 **Compare the aspects of the G_2 phase of interphase and the stages of mitosis. See Figure 8.7 for help.**

Stages	The Arrangement and Behavior of the Chromosomes	The Structure and Functions of the Mitotic Spindle	Other Cellular Details
G_2 interphase: preparing for mitosis			
Prophase			
Metaphase			
Anaphase			
Telophase			

TABLE 8.3 **Compare the aspects of meiosis I and meiosis II. See Figure 8.15 for help.**

Stage of Meiosis	How Are the Chromosomes Arranged during Metaphase? (Draw the arrangement of two tetrads)	What Separates during Anaphase: Homologous Chromosomes or Sister Chromatids?	Where Does Crossing Over Begin?
Meiosis I			
Meiosis II			

Content Quiz

Directions: Identify the *one* best answer for the multiple-choice questions. For true/false questions, determine if the statement is true or false. If false, change the underlined word(s) to make the statement true. Finally, add the correct word(s) to the fill-in-the-blank questions to make the statements true.

Biology and Society: Rain Forest Rescue

1. Cell division:
   A. produces eggs and sperm.
   B. is used in sexual reproduction.
   C. is used in asexual reproduction.
   D. is used during embryonic development.
   E. All of the above are true.

2. True or False? Many sexually reproducing plants can be induced to reproduce <u>asexually</u> in the laboratory.

3. Whether by sexual or asexual means, ____________ is at the heart of organismal reproduction.

What Cell Reproduction Accomplishes

4. Daughter cells formed during typical cell division have:
   A. identical sets of chromosomes with identical genes.
   B. different sets of chromosomes with identical genes.
   C. identical sets of chromosomes with different genes.
   D. different sets of chromosomes with different genes.

5. Sexual reproduction:
   A. uses only mitosis.
   B. involves the production of daughter cells with double the genetic material of the parent cell.
   C. uses only ordinary cell division.
   D. requires the fertilization of an egg by a sperm.
   E. is the normal process by which the cells of most organisms divide.

6. True or False? Growth of an organism and maintenance of the body are the main roles of <u>meiosis</u>.

7. True or False? A sperm or egg has <u>twice</u> as many chromosomes as its parent cell.

8. Before a cell divides, it duplicates its ____________.

9. In ____________ reproduction, the offspring and parent have identical genes.

The Cell Cycle and Mitosis

EUKARYOTIC CHROMOSOMES

10. Which one of the following statements about chromosomes is *false*?
    A. Sister chromatids separate from each other when a cell divides.
    B. The number of chromosomes in a eukaryotic cell depends upon the species.
    C. Chromosomes are made of a combination of DNA, carbohydrate, and lipid molecules.
    D. Human body cells each typically have 46 chromosomes.
    E. Before a cell begins the division process, it duplicates all of its chromosomes.

11. True or False? The <u>lipids</u> in chromosomes help control the activity of the genes.

12. Sister chromatids remain attached at a region called the ____________.

THE CELL CYCLE

13. What stage of the cell cycle is marked by the presence of sister chromatids and the preparation of the cell for division?

    A. S

    B. M

    C. G_1

    D. G_2

14. True or False? Some highly specialized cells <u>do not</u> undergo a cell cycle.

15. True or False? <u>Interphase</u> typically lasts for most of a cell cycle.

16. The M phase of the cell cycle includes ____________ and ____________.

17. During the ____________ stage of the cell cycle, the chromosomes are duplicated.

MITOSIS AND CYTOKINESIS

18. Which one of the following statements about mitosis is *false*?

    A. During prophase, the sister chromatids are clearly seen, the nuclear envelope breaks up, and the mitotic spindle begins to form.

    B. Toward the end of prophase, the sister chromatids replicate again, but remain attached to each other.

    C. During metaphase, the chromosomes line up in the middle of the mitotic spindle.

    D. At the start of anaphase, the sister chromatids are pulled apart and start to move toward opposite spindle poles.

    E. Telophase is like the opposite of prophase: The nuclear envelope re-forms, the chromosomes uncoil, and the spindle disappears.

19. Dividing an animal cell into two cells is most like:

    A. building a new wall that divides one room into two.

    B. overtightening the drawstring of sweatpants.

    C. peeling an apple.

    D. cutting a pie into two pieces with a knife.

20. Examine the prophase stages of mitosis in Figure 8.7 of your text. Which of the following structures invade(s) the nuclear space after the nuclear membrane disintegrates?

    A. chromosomes

    B. centrioles

    C. nucleolus

    D. spindle microtubules

21. Compare the cell stages in Figure 8.7 of the text to the photograph in Figure 8.1a. What stage is indicated by the cell in Figure 8.1a?
    A. interphase
    B. prophase
    C. metaphase
    D. anaphase
    E. telophase

22. True or False? The centrosomes of <u>plant</u> cells contain centrioles.

23. In plants, cytokinesis involves the formation of a membranous disk called the ____________, which helps build a new cell wall between the daughter cells.

24. Spindle microtubules grow from ____________.

CANCER CELLS: GROWING OUT OF CONTROL

25. The most dangerous property of cancer cells is that they:
    A. divide slowly.
    B. spread throughout the body.
    C. form tumors.
    D. do not divide by mitosis.
    E. do not live very long.

26. Which one of the following can *reduce* your risk of developing cancer?
    A. smoking
    B. prolonged exposure to the sun
    C. regular exercise
    D. a low-fiber, high-fat diet

27. True or False? Radiation and chemotherapy fight cancer by <u>stopping cell division</u>.

28. An abnormal mass of cells that remains at its original site is a ____________ tumor.

29. The use of drugs to disrupt cancer cell division is called ____________.

Meiosis, the Basis of Sexual Reproduction

HOMOLOGOUS CHROMOSOMES

30. A male human somatic cell has:
    A. 46 pairs of chromosomes.
    B. 23 pairs of sex chromosomes.
    C. 23 pairs of autosomes.
    D. 22 pairs of autosomes and 1 pair of sex chromosomes.
    E. 22 pairs of sex chromosomes and 1 pair of autosomes.

31. True or False? Most of the chromosomes in humans are <u>sex chromosomes</u>.

32. The same sequence of genes is found on ____________ chromosomes.

GAMETES AND THE LIFE CYCLE OF A SEXUAL ORGANISM

33. In the human life cycle, somatic cells are:
    A. diploid, and gametes are diploid.
    B. diploid, and gametes are haploid.
    C. haploid, and gametes are diploid.
    D. haploid, and gametes are haploid.

34. True or False? If the human reproductive cycle only included diploid cells, the offspring of each generation would have <u>twice as much</u> genetic material in each cell as the parent cells.

35. If the somatic cells of a species have 25 pairs of homologous chromosomes, the number of chromosomes in haploid gametes would be ____________.

THE PROCESS OF MEIOSIS

36. Which one of the following statements about meiosis is *false*?
    A. Cells dividing by meiosis undergo two consecutive divisions.
    B. Homologous chromosomes exchange segments before separating from each other.
    C. Homologous chromosomes separate during meiosis I.
    D. Sister chromatids separate during meiosis II.
    E. The chromosomes replicate before meiosis and between meiosis I and meiosis II.

37. True or False? Before crossing over occurs, sister chromatids are <u>identical</u>.

38. The result of meiosis is four ____________ cells.

REVIEW: COMPARING MITOSIS AND MEIOSIS

39. In sexually reproducing organisms, cells entering mitosis have:
    A. half the amount of DNA as cells entering meiosis.
    B. the same amount of DNA as cells entering meiosis.
    C. twice the amount of DNA as cells entering meiosis.
    D. four times the amount of DNA as cells entering meiosis.

40. Which one of the following is a true difference between mitosis and meiosis?
    A. A single cell is divided into two cells in mitosis and four cells in meiosis.
    B. Mitosis produces haploid cells, and meiosis produces diploid cells.
    C. Mitosis involves two cellular divisions, and meiosis involves just one cellular division.
    D. The chromosomes replicate before mitosis and meiosis, but in meiosis, they replicate again between the first and second division.

41. True or False? The two daughter cells produced by meiosis I <u>do not</u> undergo cytokinesis before meiosis II.

42. Sister chromatids are separated during mitosis and ____________.

THE ORIGINS OF GENETIC VARIATION, THE PROCESS OF SCIENCE: DO ALL ANIMALS HAVE SEX?

43. The random segregation of one member of each homologous pair of chromosomes into gametes defines:
    A. random fertilization.
    B. crossing over.
    C. independent assortment.
    D. random assortment.
    E. independent fertilization.

44. True or False? Crossing over only occurs in <u>mitosis</u>.

45. True or False? Animals that only reproduce <u>asexually</u> have homologous genes that differ much more from each other than homologous genes in <u>sexually</u> reproducing species.

46. The site of crossing over is a called a(n) ____________.

47. Just because of independent assortment, a man can produce nearly <u>8 million</u> different sperm.

WHEN MEIOSIS GOES AWRY

48. In general:
    A. the absence of a Y chromosome results in maleness.
    B. the presence of only a single X chromosome results in maleness.
    C. a human embryo with an abnormal number of chromosomes is usually normal.
    D. nondisjunction occurs only in males of sexually reproducing, diploid organisms.
    E. the incidence of Down syndrome increases markedly with the age of the mother.

49. Examine the chart in Figure 8.23 of your text. At which of the following intervals do we find the greatest risk of having a child with Down syndrome?
    A. 25–30 years
    B. 30–35 years
    C. 35–40 years
    D. 40–45 years
    E. 45–50 years

50. True or False? Klinefelter syndrome and Turner syndrome result from nondisjunction of <u>autosomes</u>.

51. A person with trisomy 21 is said to have ____________.

Evolution Connection: The Advantages of Sex

52. Asexual reproduction can be more adaptive when organisms are:
    A. sparsely distributed.
    B. superbly suited to a stable environment.
    C. A and B are both correct.
    D. A and B are both incorrect.

53. True or False? Asexual reproduction is more common in <u>plants</u> than in <u>animals</u>.

54. In a rapidly changing environment, ____________ reproduction may enhance survival by speeding the rate of adaptation.

Word Roots

a = not or without (asexual: type of reproduction not involving fertilization)
ana = again (anaphase: mitotic stage when sister chromatids separate)
auto = self (autosomes: the chromosomes that do not determine gender)
centro = the center; **mere** = a part (centromere: the centralized region joining two sister chromatids)
chemo = chemical (chemotherapy: type of cancer therapy using drugs that disrupt cell division)
chiasm = cross-mark (chiasma: the sites where crossing over has occurred)
chroma = colored (chromosome: DNA-containing structure)
cyto = cell; **kinet** = move (cytokinesis: division of the cytoplasm)
di = two (diploid: cells that contain two homologous sets of chromosomes)
fertil = fruitful (fertilization: process of fusion of sperm and egg cell)
gamet = a wife or husband (gamete: egg or sperm)
haplo = single (haploid: cells that contain only one chromosome of each homologous pair)
homo = like (homologous: like chromosomes that form a pair)
inter = between (interphase: time when a cell metabolizes and performs its various functions)
karyo = nucleus (karyotype: a display of the chromosomes of a cell)
mal = bad or evil (malignant: type of tumor that migrates away from its site of origin)
mei = less (meiosis: the division of a diploid nucleus into four haploid daughter nuclei)
meta = between (metaphase: mitotic stage when the chromosomes are lined up in the cell's middle)
mito = a thread (mitosis: the division of a diploid cell into two diploid cells)
non = not; **dis** = separate (nondisjunction: the result when paired chromosomes fail to separate)
pro = before (prophase: mitotic stage when the nuclear membrane first breaks up)
soma = body (somatic: body cells with 46 chromosomes in humans)
telo = end (telophase: final mitotic stage when the nuclear envelope re-forms)
tri = three (trisomy 21: a condition in which a person has three number 21 chromosomes)

Key Terms

anaphase
asexual reproduction
autosome
benign tumor
cancer
cell cycle
cell cycle control system
cell division
cell plate
centromere
centrosome
chemotherapy
chiasma
chromatin
chromosome
cleavage furrow
crossing over
cytokinesis
diploid
Down syndrome
fertilization
gamete
genetic recombination
haploid
histone
homologous chromosome
interphase
karyotype
life cycle
malignant tumor
meiosis
metaphase
metastasis
mitosis
mitotic phase
mitotic spindle
nucleosome
nondisjunction
prophase
radiation therapy
sex chromosome
sexual reproduction
sister chromatid
somatic cell
telophase
trisomy 21
tumor
zygote

Crossword Puzzle

Use the Key Terms list from this chapter to fill in the crossword puzzle.

ACROSS

1. type of tumor that stays at its original site of origin
4. sequence of events from cell formation to cell division
8. disease of cells that divide excessively and exhibit bizarre behavior
10. process of fusion of haploid sperm and haploid egg cell
11. type of reproduction not involving fertilization
13. division of the cytoplasm
14. the region where two chromatids are joined
15. cells that contain only one chromosome of each homologous pair
17. mitotic stage when sister chromatids separate
19. a person with trisomy 21
20. the use of drugs to disrupt the division of cancer cells
21. the chromosomes that do not determine sex
23. exchange of genetic material between homologous chromosomes
24. small protein associated with DNA
27. a thread-like, gene-carrying structure formed from chromatin
29. a condition in which a person has three number 21 chromosomes
30. the sites where crossing over has occurred
32. football-shaped structure of microtubules important in mitosis
33. stages leading from the adults of one generation to the adults of the next
35. a display of the chromosomes of a cell
36. an abnormally growing mass of body cells
37. type of tumor that migrates away from its site of origin
38. mitotic stage when the nuclear membrane first breaks up

DOWN

2. production of gene combinations unlike the parent chromosomes
3. cell reproduction
5. membranous disk containing cell wall material in plants
6. the fertilized egg
7. long strands of DNA attached to certain proteins
9. cells that contain two homologous sets of chromosomes
12. indentation at the equator of a cell where it will divide
14. clouds of cytoplasmic material with centrioles in animal cells
16. the use of drugs to disrupt the division of cancer cells
18. egg or sperm
22. the division of a diploid nucleus into four haploid daughter nuclei
25. one of two identical parts of a replicated chromosome
26. type of cancer therapy exposing parts of the body to high energy
28. type of chromosome that forms a pair
31. mitotic stage when the chromosomes are lined up in the cell's middle
34. the result when paired chromosomes fail to separate
40. the division of a diploid cell into two diploid cells
43. type of reproduction requiring fertilization
44. type of chromosome that determines the sex of the child

ACROSS

39. type of cell with 46 chromosomes in humans
41. time when a cell metabolizes and performs its various functions
42. the spread of cancer cells beyond their original site
45. final mitotic stage when the nuclear envelope re-forms
46. part of cell cycle when cell is dividing
47. a bead-like structure consisting of DNA wrapped around histone molecules

CHAPTER 9

Patterns of Inheritance

Studying Advice

a. Have you ever wondered how your sex was determined, why people have different skin colors, and how diseases such as hemophilia and sickle-cell anemia are inherited? This chapter addresses some of the most interesting questions and relevant information in all of your biological studies!

b. This chapter includes many examples of different types of inheritance. To help you organize these definitions, Table 9.2 provides space for you to define and describe these key terms/phrases related to inheritance.

c. If you have not recently studied Chapter 8, you will need to review the process of meiosis to best understand the Chapter 9 section titled "The Chromosomal Basis of Inheritance."

Student Media

Activities

Monohybrid Cross
Dihybrid Cross
Gregor's Garden
Incomplete Dominance
Linked Genes and Crossing Over
Sex-Linked Genes

Biology Labs On-Line

PedigreeLab
FlyLab

BLAST Animations

Single-Trait Crosses
Genetic Variation: Independent Assortment
Two-Trait Crosses

LabBench

Genetics of Organisms

MP3 Tutor

Chromosomal Basis of Inheritance

Process of Science

What Can Fruit Flies Reveal about Inheritance?

Videos

Discovery Channel Video: Colored Cotton
Discovery Channel Video: Novelty Gene
Ultrasound of Human Fetus 1

Organizing Tables

TABLE 9.1 **The table shows the results of a cross between plants heterozygous for purple flower color. Write the genotypes of the plants in each square as either (a) homozygous dominant, (b) heterozygous, or (c) homozygous recessive. Next, write the phenotypes of the flowers for each square as either (a) purple (the dominant trait) or (b) white (the recessive trait).**

Genotype: ***PP***	**Genotype:** ***Pp***
Genotype in words ____________ Phenotype ____________	Genotype in words ____________ Phenotype ____________
Genotype: ***pP***	**Genotype:** ***pp***
Genotype in words ____________ Phenotype ____________	Genotype in words ____________ Phenotype ____________

TABLE 9.2 Key terms: This table should help you organize the definitions of key aspects of inheritance.

Term	Definition
Recessive disorders	
Dominant disorders	
Incomplete dominance	
Multiple-allele inheritance	
Codominance	
Pleiotropy	
Polygenic inheritance	
Linked genes	
Sex-linked genes	

Content Quiz

Directions: Identify the *one* best answer for the multiple-choice questions. For true/false questions, determine if the statement is true or false. If false, change the underlined word(s) to make the statement true. Finally, add the correct word(s) to the fill-in-the-blank questions to make the statements true.

Biology and Society: A Matter of Breeding

1. Crossing two dogs of the same pedigree will most likely produce:
   A. mutts.
   B. animals of diverse pedigrees.
   C. offspring with a different pedigree.
   D. offspring with a similar pedigree.

2. True or False? Generations of <u>inbreeding</u> purebred dogs has led to serious genetic defects.

3. The field of ____________ is the scientific study of heredity.

Heritable Variation and Patterns of Inheritance

IN AN ABBEY GARDEN

4. Crossing members of two different true-breeding varieties of organisms produces:
   A. F_1 hybrids.
   B. F_2 hybrids.
   C. P_1 hybrids.
   D. P_2 hybrids.
   E. F_1 true breeders.

5. True or False? When Mendel wanted to <u>cross-fertilize</u> his pea plants, he covered the flower with a small bag.

6. Fertilization between members of the F_1 generation produces members of the ______________ generation.

MENDEL'S LAW OF SEGREGATION

7. What are the phenotypes of two pea plants with the flower color genotypes Pp and PP?
   A. Both are purple.
   B. Both are white.
   C. One is white and the other is purple.
   D. A Punnett square is needed to figure this out.

8. If we crossed two pea plants with genotypes Pp and PP as noted in question 7, we would expect that the phenotypes of the offspring would be:
   A. all purple.
   B. half purple and half white.
   C. three quarters purple and one quarter white.
   D. three quarters white and one quarter purple.
   E. all white.

9. Mendel's principle of segregation states that:
   A. pairs of alleles stay together during gamete formation and the alleles separate at fertilization.
   B. pairs of alleles segregate during gamete formation and the alleles separate at fertilization.
   C. pairs of alleles stay together during gamete formation and alleles pair again at fertilization.
   D. pairs of alleles segregate during gamete formation and alleles pair again at fertilization.

10. True or False? An individual who has two different alleles for a gene is said to be <u>homozygous</u>.

11. The physical location of a gene on a chromosome is that gene's ______________.

MENDEL'S LAW OF INDEPENDENT ASSORTMENT

12. If Mendel's principle of independent assortment did not apply to the pea shape and color experiment, and the two traits in a dihybrid cross of heterozygous individuals were inherited together, what would be the expected phenotypic ratio of the F_2 generation?
    A. 9:3:3:1
    B. 1:6:1
    C. 1:2:1
    D. 3:1
    E. 1:1

13. True or False? Mendel performed dihybrid crosses involving all seven of his pea characters and found 9:3:3:1 ratios in half of them.

14. Mendel's pea shape and color experiment showed that this dihybrid cross was the equivalent of two ____________ crosses occurring simultaneously.

USING A TESTCROSS TO DETERMINE AN UNKNOWN GENOTYPE

15. A testcross is a mating between:
    A. an organism of unknown genotype and a homozygous dominant individual.
    B. an organism of unknown genotype and a heterozygous individual.
    C. an organism of unknown genotype and a homozygous recessive individual.
    D. two heterozygous individuals.
    E. a homozygous dominant and a homozygous recessive individual.

16. True or False? Mendel used testcrosses to determine whether he had true-breeding varieties of plants.

17. Testcrosses are used to determine the ____________ of an organism.

THE RULES OF PROBABILITY

18. Using a standard deck of 52 playing cards, the chance of drawing a red card is 1/2 and the chance of drawing a queen is 1/13. To determine the probability of drawing a red queen, we should use the rule of:
    A. addition.
    B. subtraction.
    C. multiplication.
    D. division.

19. True or False? If we know the phenotypes of the parents, we can predict the probability for any genotype among the offspring.

FAMILY PEDIGREES

20. Consider a trait in which "T" represents the dominant and "t" represents the recessive alleles. Which of the following genotypes will exhibit the recessive phenotype?

    A. TT

    B. Tt

    C. tT

    D. tt

    E. More than one of the above.

    F. None of the above.

21. Examine the patterns of inheritance of attached earlobes in Figure 9.13 of your text. How many of the six members of the second generation are carriers for the attached earlobes trait?

    A. 0

    B. 2

    C. 5

    D. 7

22. True or False? Dominant phenotypes <u>are</u> more common than recessive phenotypes.

23. True or False? A family <u>pedigree</u> is used to assemble information about the occurrence of heritable characters in parents and their offspring across several generations.

24. People who are heterozygous with just one copy of an allele for a recessive disorder do not show symptoms of the disorder. They are considered to be _____________ of the disorder.

HUMAN DISORDERS CONTROLLED BY A SINGLE GENE

25. Dominant lethal alleles:

    A. are more common than lethal recessive alleles.

    B. are commonly carried by heterozygotes without affecting them.

    C. are never inherited from a heterozygote.

    D. cause most human genetic disorders.

    E. cause Huntington's disease.

26. True or False? Inbreeding is <u>less</u> likely to produce offspring that are homozygous for a harmful recessive trait.

27. Most people born with recessive disorders are born to parents who are both _____________ or carriers for the recessive allele.

28. The illness called _____________ is a degeneration of the nervous system that usually does not begin until middle age.

29. The most common lethal genetic disease in the United States is _____________.

THE PROCESS OF SCIENCE: WHAT IS THE GENETIC BASIS OF HAIRLESS DOGS?

30. The hairless condition in dogs is a result of:
    A. a mutation in a single gene.
    B. shedding too frequently.
    C. two new alleles.
    D. the loss of a chromosome.
    E. incomplete meiosis.

31. True or False? The hairless phenotype is inherited as a <u>dominant</u> trait.

32. The ____________ dominant condition of the hairless trait is lethal.

Variations on Mendel's Laws

INCOMPLETE DOMINANCE IN PLANTS AND PEOPLE

33. If a mating occurs between two parent plants that are both heterozygous for a trait with incomplete dominance, the expected ratio of phenotypes will be:
    A. 1:3.
    B. 1:2:1.
    C. 1:1.
    D. 9:3:3:1.
    E. 2:1.

34. True or False? Mendel's work involved plants that <u>showed</u> incomplete dominance.

35. Humans with the disease ____________ are homozygous or heterozygous for an allele that causes dangerously high cholesterol levels in the blood.

ABO BLOOD GROUPS: AN EXAMPLE OF MULTIPLE ALLELES AND CODOMINANCE

36. What is the expected <u>phenotypic</u> ratio of children from parents with the following blood genotypes? I^A I^A and I^B I^i?
    A. 1/2 AB and 1/2 A
    B. 1/2 AB and 1/2 B
    C. 1/4 A, 1/2 AB, and 1/4 B
    D. 1/2 AB and 1/2 O
    E. 1/4 B, 1/2 A, and 1/4 O

37. Examine the blood test results in Figure 9.20 of your text. Why aren't any of the blood cells clumped in the test of AB blood?
    A. There must have been a mistake.
    B. AB blood has antibodies to types A and B carbohydrates.
    C. AB blood does not have antibodies to type A or B carbohydrates.
    D. AB blood has antibodies to type A carbohydrates.
    E. AB blood has antibodies to type B carbohydrates.

38. True or False? When a trait exhibits codominance, both recessive phenotypes are expressed in heterozygotes with two dominant alleles.

39. People with type ____________ blood show codominance because both alleles are expressed.

PLEIOTROPY AND SICKLE-CELL DISEASE

40. Which one of the following is an example of pleiotropy in humans?
    A. hypercholesterolemia
    B. achondroplasia
    C. Huntington's disease
    D. sickle-cell disease
    E. cystic fibrosis

41. True or False? The direct effect of the sickle-cell allele is to make red blood cells produce abnormal hemoglobin proteins.

42. At the molecular level, the alleles of sickle-cell disease are ____________ because both are expressed in heterozygous individuals.

POLYGENIC INHERITANCE

43. In the example of polygenic inheritance described in the text, regarding human skin color, which of the following genotypes would produce the darkest skin?
    A. AAbbcc
    B. AaBbCc
    C. aabbcc
    D. AAbBcc
    E. AaBbCC

44. True or False? In polygenic inheritance of skin color, only the dominant alleles contribute to darker skin color.

45. The opposite of pleiotropy, ____________ involves two or more genes affecting a single character.

THE ROLE OF ENVIRONMENT

46. If we examine a real human population for the skin color phenotype, we would see more shades than just seven because:
    A. of the effects of environmental factors.
    B. some genes fade over time, producing light effects.
    C. there appears to be more genes involved in skin tone than have been described.
    D. skin color has no genetic component.

47. True or False? Susceptibility to heart disease clearly has an environmental component.

48. Although organisms result from a combination of genetic and environmental factors, only ____________ factors are passed on to the next generation.

The Chromosomal Basis of Inheritance

49. Linked genes tend to be inherited together because:
    A. they affect the same character.
    B. they determine different aspects of the same trait.
    C. they have similar structures.
    D. they are on the same chromosome.
    E. they have the same alleles.

50. Which one of the following processes can result in the separate inheritance of linked genes?
    A. polygenic inheritance
    B. crossing over
    C. incomplete dominance
    D. pleiotropy
    E. independent assortment

51. The probability of crossover between two linked genes is greatest when the genes are:
    A. located on separate chromosomes.
    B. close together on the same chromosome.
    C. farthest apart on the same chromosome.
    D. pleiotropic.
    E. dominant.

52. True or False? Linked genes <u>do not</u> follow the typical patterns of inheritance.

53. True or False? Researchers used recombination data to assign genes to relative positions on chromosomes to create <u>chromosome</u> maps.

54. The ____________ states that genes are located at specific positions on chromosomes and that chromosomal behavior during meiosis and fertilization accounts for inheritance patterns.

55. Early studies of the relationship between chromosome behavior and inheritance relied upon ____________, often seen flying around overripe fruit.

Sex Chromosomes and Sex-Linked Genes

56. Which one of the following statements about sex chromosomes is <u>false</u>?
    A. Humans have 44 autosomes and 2 sex chromosomes.
    B. In humans, a male inherits an X and Y sex chromosome.
    C. Most sex-linked genes unrelated to sex determination are found on the X chromosome.
    D. Each human gamete contains both sex chromosomes.

57. For a sex-linked recessive allele to be expressed, a man would have to inherit:
    A. only one recessive allele, but a woman would have to inherit two.
    B. two recessive alleles, but a woman would have to inherit only one.
    C. two recessive alleles, just like a woman.
    D. only one recessive allele, just like a woman.

58. True or False? Any gene located on a sex chromosome is a <u>sex-linked gene</u>.

59. True or False? Sex-linked human diseases <u>do not occur</u> in females.

60. True or False? Red-green colorblindness is a common sex-linked disorder involving <u>one</u> X-linked gene(s).

61. The ____________ chromosomes of a human female are XX.

62. Factors involved in blood clotting are missing in people suffering from ____________.

Evolution Connection: Barking Up the Evolutionary Tree

63. True or False? The many existing breeds of dogs are all the result of <u>natural</u> selection.

64. About 15,000 years ago, canines existed that are the ancestors of modern ____________ and dogs.

Word Roots

co = together (codominance: phenotype in which both dominant alleles in a heterozygous individual are expressed)
di = two (dihybrid: a type of cross that mates varieties differing in two characters)
geno = offspring (genotype: an organism's genetic makeup)
hemo = blood; **philia** = love (hemophilia: a human genetic disease caused by excessive bleeding following an injury)
hetero = different (heterozygous: when an organism has different alleles for a gene)
homo = alike (homozygous: when an organism has the same alleles for a gene)
hyper = excessive (hypercholesterolemia: a condition of incomplete dominance resulting in elevated blood cholesterol levels)
mono = one (monohybrid: a type of cross between organisms that differ in only one trait)
pheno = appear (phenotype: an organism's physical traits)
pleio = more; **trop** = change (pleiotropy: when a single gene impacts more than one character)
poly = many; **gen** = produce (polygenic: type of inheritance in which two or more genes affect a single trait)

Key Terms

ABO blood groups
achondroplasia
alleles
carrier
character
chromosome theory of inheritance
codominance
cross
dihybrid cross
dominant allele
F_1 generation
F_2 generation
genetics
genotype
hemophilia
heredity
heterozygous
homozygous
Huntington's disease
hybrid
hypercholesterolemia
inbreeding
incomplete dominance
law of independent assortment
law of segregation
linkage map
linked genes
loci
monohybrid cross
P generation
pedigree
phenotype
pleiotropy
polygenic inheritance
Punnett square
recessive allele
recombination frequency
red-green color blindness
rule of multiplication
sex-linked gene
sickle-cell disease
testcross
trait
true-breeding
wild-type traits

Crossword Puzzle

Use the Key Terms list from this chapter to fill in the crossword puzzle.

ACROSS

2. when an organism has the same alleles for a gene
3. a device for predicting the results of a genetic cross
4. when an organism has different alleles for a gene
5. the transmission of traits from one generation to the next
9. the type of traits most often seen in nature
13. hybridization of F1 organisms
15. the genetic makeup of an organism
17. the offspring of two different true-breeding varieties
18. the specific locations of genes on chromosomes
19. a type of map based on the frequencies of recombinations during crossover
22. type of inheritance in which two or more genes affect a single trait
23. phenotype in which both dominant alleles in a heterozygous individual are expressed
25. a heritable feature that varies among individuals
26. the three letters representing the three types of humans blood alleles
27. results of a cross of close relatives
28. alternate forms of genes
30. a hybridization
31. type of genes that are located close together on the same chromosome

32. in a heterozygote, the type of allele that determines the phenotype
33. a type of cross between organisms that differ in only one trait
35. type of dominance when the heterozygote phenotype is intermediate to the homozygous phenotypes
36. an organism's physical traits
39. the principle that sperm or eggs carry only one allele for each inherited character
40. a variant of a character
41. a type of colorblindness commonly sex-linked in humans
42. in a heterozygote, the type of allele that has no noticeable affect on the phenotype
43. when a single gene impacts more than one character
44. the percentage of recombinant offspring in a testcross

DOWN

1. production of offspring with inherited trait(s) identical to those of the parents
2. a sex-linked recessive trait that causes excessive bleeding
6. the way to determine the probability that two independent events will both occur
7. a type of disease of malformed blood cells and other symptoms
8. hybrid offspring
10. the science of heredity
11. the type of law that states that each pair of alleles sorts independently of the other pairs of alleles during gamete production
12. a type of degenerative disease of the nervous system caused by a dominant allele
14. a type of cross mating varieties differing in two characters
16. a condition of incomplete dominance resulting in elevated blood cholesterol levels
20. an organism that has one allele for a recessive disorder but shows no symptoms
21. a form of dwarfism caused by a dominant allele
24. the type of theory of inheritance that states that genes are located on specific positions on chromosomes
29. any gene located on a sex chromosome
34. mating of an organism of unknown genotype with a homozygous recessive organism
37. family tree describing the occurrence of heritable characters in parents and offspring
38. the parental organisms

CHAPTER 10

The Structure and Function of DNA

Studying Advice

a. This chapter introduces the basics of molecular genetics, one of the most exciting and explosive fields of modern biology. A mastery of this chapter's content is necessary to understand the discussions in Chapters 11 and 12.

b. Before reading this chapter, review the structure of nucleic acids in Chapter 3.

c. Begin studying this chapter by examining Figure 10.8. This figure represents the overall flow of genetic information in a cell and is fundamental to the rest of the chapter.

Student Media

Activities

The Hershey-Chase Experiment
DNA and RNA Structure
DNA Double Helix
DNA Replication: An Overview
Overview of Protein Synthesis
Transcription
RNA Processing
Translation
Simplified Viral Reproductive Cycle
Phage Lytic Cycle
Phage Lysogenic and Lytic Cycles
Retrovirus (HIV) Reproductive Cycle

BioFlix

DNA Replication
Protein Synthesis

Biology Labs On-Line

TranslationLab

BLAST Animations

Hydrogen Bonds in DNA
Structure of DNA Double Helix
Transcription
Translation
Roles of RNA
HIV Structure
AIDS Treatment Strategies

MP3 Tutors

DNA to RNA to Protein

Process of Science

What Is the Correct Model for DNA Replication?
How Is a Metabolic Pathway Analyzed?
How Do You Diagnose a Genetic Disorder?
What Causes Infections in AIDS Patients?
Why Do AIDS Rates Differ Across the U.S.?

Videos

Discovery Channel Video: Vaccines
Discovery Channel Video: Emerging Diseases

Organizing Tables

TABLE 10.1 Compare the structure of DNA and mRNA in the table.

	Type of Sugar	Types of Bases	Overall Shape of the Molecule
DNA			
mRNA			

TABLE 10.2 Compare the structures and functions of the different types of RNA molecules in the table.

	Structure	Function	Location in the Cell
mRNA			
tRNA			
rRNA			

TABLE 10.3 Compare the different types of infectious agents in the table.

	Structure	How It Reproduces
Bacteriophages		lytic cycle lysogenic cycle
Retroviruses		
Viroids		
Prions		

Content Quiz

Directions: Identify the *one* best answer for the multiple-choice questions. For true/false questions, determine if the statement is true or false. If false, change the underlined word(s) to make the statement true. Finally, add the correct word(s) to the fill-in-the-blank questions to make the statements true.

Biology and Society: Tracking a Killer

1. Flu viruses consist of:
   A. DNA and RNA.
   B. RNA and protein.
   C. DNA and protein.
   D. just DNA.
   E. just RNA.

2. About how many people die each year in the United States from influenza infection?
   A. 1,000
   B. 5,000
   C. 20,000
   D. 100,000
   E. 500,000

3. True or False? The flu is caused by a <u>bacterium</u>.

DNA: Structure and Replication

4. In the 1950s, scientists understood the functions of DNA to be all of the following *except* the capacity to:
   A. *store* genetic information.
   B. *create* new genetic information.
   C. *copy* genetic information.
   D. *pass along* genetic information from generation to generation.

5. True or False? Mendel worked on inheritance patterns <u>without knowing</u> about DNA's role in heredity.

6. By the 1950s, a race was on to understand the ____________ of DNA.

DNA AND RNA STRUCTURE

7. The backbone of DNA and RNA, polynucleotides consist of a repeating pattern of:
   A. sugar, base, sugar, base.
   B. phosphate, base, phosphate, base.
   C. sugar, phosphate, sugar, phosphate.
   D. sugar, base, phosphate, sugar, base, phosphate.

8. True or False? The "D" in DNA comes from deoxyribose because compared to the sugar in RNA, the sugar in DNA <u>has an extra</u> oxygen atom.

9. The sugars of DNA and RNA are different. DNA has the sugar ____________, and RNA has the sugar ____________.

10. RNA uses the nitrogenous base ____________, which DNA does not use.

11. The two types of bases in DNA are the single-ring bases, ____________ and ____________, and the double-ring structures, ____________ and ____________.

12. In DNA and RNA, the polymers are ____________ and the monomers are ____________.

WATSON AND CRICK'S DISCOVERY OF THE DOUBLE HELIX

13. The genetic information in a chromosome is encoded:
    A. in the nucleotide sequence of the molecule.
    B. by the base pairing of the DNA molecule.
    C. by the interaction between DNA and the associated histone proteins.
    D. in the types of sugars used in the molecule.
    E. in the types of chemical bonds formed between the bases forming DNA.

14. The shape of a DNA molecule is most like the shape of:
    A. a vase.
    B. a spiral staircase.
    C. the strings on a tennis racquet.
    D. the letter *X*.

15. If adenine paired with guanine and cytosine paired with thymine in DNA, then:
    A. DNA would have irregular widths along its length.
    B. the DNA molecule would be much longer.
    C. the DNA molecule would be much shorter.
    D. the sequential information would be lost.
    E. the DNA molecule would be circular.

16. True or False? Watson and Crick discovered that the backbone of DNA was located on the <u>inside</u> of the molecule.

17. If one side of a DNA molecule has the bases CGAT, the opposite side would have the bases ____________.

DNA REPLICATION

18. When a DNA molecule is copied, how much of the original DNA is included in the new copy?
    A. none
    B. 25%
    C. 50%
    D. 75%
    E. 100%

19. DNA replication is most like:
    A. picking students to make two baseball teams.
    B. two people getting divorced followed by each person remarrying.
    C. splitting a large plant into two and planting each half.
    D. mixing vinegar and oil to make salad dressing.

20. True or False? The DNA molecule of a eukaryotic chromosome has <u>a single replication origin</u>.

21. Covalent bonds between the nucleotides of a new DNA strand are made by the enzymes ____________.

22. DNA repair is accomplished by the enzymes ____________ and some of the proteins associated with DNA replication.

The Flow of Genetic Information from DNA to RNA to Protein

HOW AN ORGANISM'S GENOTYPE DETERMINES ITS PHENOTYPE

23. The Beadle and Tatum hypothesis about the function of genes is now revised as one gene–:
    A. one protein.
    B. one enzyme.
    C. one DNA.
    D. one polypeptide.
    E. one monomer.

24. Which one of the following best represents the flow of genetic information in a cell?
    A. RNA → transcription → DNA → translation → PROTEIN
    B. DNA → transcription → RNA → translation → PROTEIN
    C. DNA → transcription → PROTEIN → translation → RNA
    D. DNA → translation → RNA → transcription → PROTEIN

25. True or False? The molecular basis of the phenotype lies in an organism's <u>DNA</u>.

26. An organism's ____________ is its genetic makeup. An organism's physical traits are its ____________.

27. Cells produce RNA by the process of ____________ and proteins by the process of ____________.

FROM NUCLEOTIDES TO AMINO ACIDS: AN OVERVIEW

28. If a very short gene consisted of just 60 bases, the protein made from the gene would be about ____________ amino acids long.
    A. 10
    B. 20
    C. 60
    D. 180

29. True or False? A DNA molecule may contain <u>thousands</u> of genes.

30. True or False? A gene may consist of <u>thousands</u> of nucleotides.

31. A codon consists of ____________ base(s) in a DNA or RNA molecule.

32. The sequence of nucleotides of the RNA molecule dictates the sequence of ____________ of the polypeptide.

THE GENETIC CODE

33. Of the 64 possible codons:
    A. 61 code for an amino acid and 3 are stop codons.
    B. all of them code for at least one amino acid.
    C. 20 code for 2 amino acids and the rest code for a single amino acid.
    D. 2 are start codons, 3 are stop codons, and 59 code for a single amino acid.
    E. 1 is a start codon, 1 is a stop codon, and the remaining 62 code for a single amino acid.

34. Examine the sets of codons in Figure 10.10 that code for the same amino acid. How do the codons that all code for the same amino acid compare?
    A. The first letter in the codon is most likely to be different.
    B. The second letter in the codon is most likely to be different.
    C. The third letter in the codon is most likely to be different.
    D. The second and third letters in the codon are always the same.
    E. The first and third letters in the codon are always the same.

35. True or False? There are <u>gaps</u> between the codons of RNA.

36. True or False? Different codons may code <u>for the same amino acid</u>.

37. The set of rules relating nucleotide sequence to amino acid sequence is called the ____________.

TRANSCRIPTION: FROM DNA TO RNA

38. Which one of the following statements about transcription is *false*?
    A. In RNA, U, rather than T, pairs with A.
    B. The RNA molecule is built one nucleotide at a time.
    C. Both DNA strands serve as the template for one RNA.
    D. Transcription begins when RNA polymerase attaches to the promoter.
    E. As the RNA molecule is produced, it peels away from its DNA template.

39. True or False? The terminator sequence indicates the <u>end</u> of a gene.

40. The ____________ dictates which of the two DNA strands is to be transcribed.

41. The RNA nucleotides are linked by the enzyme ____________.

THE PROCESSING OF EUKARYOTIC RNA

42. In eukaryotic cells, RNA transcribed from DNA undergoes processing before leaving the nucleus. In this processing:
    A. a cap and tail are added, introns are edited out, and exons are joined together to make mRNA.
    B. a cap and tail are removed, introns are edited out, and exons are joined together to make mRNA.
    C. a cap and tail are added, exons are edited out, and introns are joined together to make mRNA.
    D. a cap is removed, a tail is added, exons are edited out, and introns are joined together to make mRNA.

43. True or False? The cap and tail on an mRNA molecule <u>protect mRNA from cellular enzymes</u> and help ribosomes recognize RNA as mRNA.

44. RNA splicing involves the removal of ____________ and the joining of ____________ to produce mRNA.

TRANSLATION: THE PLAYERS

45. The actual translation of codons into amino acids is the job of:
    A. mRNA.
    B. tRNA.
    C. rRNA.
    D. ribosomes.
    E. RNA polymerase.

46. Which one of the following statements about tRNA is *false*?

A. An anticodon recognizes a particular mRNA codon by using base-pairing rules.

B. There is a slightly different version of tRNA for each amino acid.

C. Some parts of the tRNA molecule twist, fold around, and base-pair with itself.

D. Each tRNA molecule must pick up the appropriate amino acid.

E. A tRNA molecule is made of a double strand of RNA about 800 nucleotides long.

47. A ribosome consists of:

A. one subunit made up of proteins and DNA.

B. two subunits, each made up of tRNA and rRNA.

C. two subunits, each made up of proteins and mRNA.

D. two subunits, each made up of proteins and rRNA.

E. three subunits, each made up of proteins, mRNA, and tRNA.

48. True or False? The part of a tRNA molecule that binds to a codon is a(n) <u>parallel codon</u>.

49. A fully assembled ribosome has a binding site for ____________ on its small subunit and a binding site for ____________ on its large subunit.

TRANSLATION: THE PROCESS, REVIEW: DNA → RNA → PROTEIN

50. During the initiation stage of translation:

A. a ribosome assembles with the mRNA and the initiator tRNA bearing the first amino acid.

B. additional amino acids are brought in, one at a time, as a polynucleotide forms.

C. the ribosomal subunits form by the combination of rRNA and proteins.

D. the mRNA molecule is edited further, beginning with the removal of the cap and tail.

51. Which one of the following is the correct molecular sequence in the combined processes of transcription and translation?

A. DNA → mRNA → polypeptide → protein

B. mRNA → DNA → polypeptide → protein

C. DNA → polypeptide → mRNA → protein

D. DNA → mRNA → protein → polypeptide

E. mRNA → polypeptide → DNA → protein

52. True or False? An incoming tRNA molecule first binds with the mRNA codon <u>at the A site</u>.

53. After a new amino acid is brought in, the polypeptide leaves the tRNA in the P site and attaches to the ____________ on the tRNA in the A site.

54. Once a new amino acid is added to a polypeptide, the tRNA and mRNA move together from the ____________ site to the ____________ site. This process is known as ____________.

55. Elongation continues until a(n) ____________ reaches the ribosome's A site.

MUTATIONS

56. Compare the two sentences below.

The cat hit the red toy pig.

The caz thi tth ere dto ypi.

The second "sentence" has been changed in a way that is most like a mutation caused by:

A. a base substitution.

B. a single base deletion.

C. a single base addition.

D. a multiple base deletion.

E. a multiple base addition.

57. In the mutation in question 56, the second sentence no longer reads properly because:

A. of a shift in the triplet grouping.

B. most of the letters are different.

C. several key letters were lost.

D. of word substitutions.

E. of word deletions.

58. True or False? The sickle-cell allele differs from its normal counterpart by <u>one</u> nucleotide.

59. True or False? A base substitution <u>shifts</u> the triplet grouping.

60. True or False? Mutations are usually <u>beneficial</u>.

61. Any change in the nucleotide sequence of DNA is a(n) ____________.

62. Chemicals and X-rays that cause mutations are called ____________.

Viruses and Other Noncellular Infectious Agents

BACTERIOPHAGES

63. In the lytic cycle of a bacteriophage:
    A. the DNA inserts by genetic recombination into the bacterial chromosome.
    B. the phage genes remain inactive.
    C. the viral DNA can be passed along through many generations of infected bacteria.
    D. the DNA immediately turns the cell into a virus-producing factory.
    E. the phage DNA is referred to as a prophage.

64. Which one of the following is most like the way that phage DNA is replicated during a lysogenic cycle?
    A. a smudge on an original document is included on the photocopies for class
    B. having your friend wash your clothes, fold them, and return them to your room
    C. asking a friend to help you while you change the tires on your truck
    D. inviting friends over to your house to share a good home-cooked meal
    E. working with a group of friends to design an experiment for a science class

65. True or False? The lysogenic cycle leads to the lysis of the host bacterial cell.

66. True or False? Once a prophage forms, it cannot leave its chromosome.

67. True or False? Prophage DNA is replicated along with the host cell's DNA.

68. A virus that attacks a bacterium is called a(n) ____________.

PLANT VIRUSES, ANIMAL VIRUSES

69. Which one of the following statements about plant viruses is *false*?
    A. Most plant viruses discovered to date have RNA instead of DNA as their genetic material.
    B. Many plant viruses consist of RNA surrounded by proteins.
    C. Plant viruses cannot easily penetrate a healthy plant epidermis.
    D. Plants infected with viruses can pass the virus on to their offspring.
    E. Unlike animals, plants are easily treated for viral diseases.

70. Which one of the following does *not* typically occur during the reproductive cycle of an enveloped RNA virus?
    A. A protein-coated RNA enters the host cell. Once inside the host cell, the protein coat around the virus is removed.
    B. A viral enzyme starts making complementary strands of the viral RNA.
    C. Some of the new RNA serves as mRNA for the synthesis of new viral proteins.
    D. The new viral proteins assemble around new viral RNA.
    E. The new viruses escape by killing the cells.

71. True or False? The mumps virus has been almost eliminated by antibiotics.

72. True or False? We usually recover from colds by replacing cells damaged by the virus.

73. The herpes virus DNA may remain as ____________ in the nuclei of certain nerve cells.

74. Most RNA viruses get their outer membranes from the ____________ of the host cell. However, DNA viruses, such as the herpes virus, get their outer membranes from the ____________ of the host cell.

HIV, THE AIDS VIRUS

75. Which one of the following statements about HIV is *false*?
    A. The virus that is transmitted into cells contains RNA.
    B. Once in a host cell, the viral RNA synthesizes more RNA from the host's DNA.
    C. Double-stranded DNA produced by reverse transcription is inserted into the host cell's chromosomal DNA as a provirus.
    D. The provirus is transcribed and translated into viral proteins.
    E. New HIV leaves without killing the host cell.

76. The outer envelope of HIV is derived from the ____________ of a previous host cell.

77. HIV is an example of a(n) ____________, a virus that reproduces by means of a DNA molecule.

78. HIV uses the enzyme ____________ to catalyze reverse transcription.

VIROIDS AND PRIONS

79. Prions cause disease by:
    A. turning the cell into a prion-producing factory.
    B. disrupting the plasma membrane of the host cell.
    C. inserting the prion DNA into chromosomes and stopping the process of transcription.
    D. converting normal proteins into misfolded proteins such as the prion.
    E. disrupting the ability of ribosomes to form and engage in translation.

80. True or False? Viroids do not encode proteins.

81. Viroids are small circular ____________ molecules and prions are infectious ____________ molecules.

Evolution Connection: Emerging Viruses

82. RNA viruses mutate more quickly than our DNA because RNA viruses:
    A. come into contact with many mutagens.
    B. are made of more fragile amino acids.
    C. are smaller.
    D. lack proofreading steps after replication.
    E. have weaker covalent bonds between their bases.

83. True or False? Annual flu vaccines are necessary because last year's flu virus <u>may be mutated enough</u> that you have little immunity against it.

84. True or False? New viral diseases may result when an old virus is introduced to a new <u>host</u>.

85. Viruses that have appeared suddenly or may have only recently come to the attention of science are called ____________ viruses.

Word Roots

muta = change; **gen** = producing (mutagen: a physical or chemical agent that causes mutations)
phage = eat (bacteriophages: viruses that attack bacteria)
poly = many (polynucleotide: a polymer of many nucleotides)
pro = before (prophage: phage DNA inserted into the bacterial chromosome before viral replication)
retro = backward (retrovirus: an RNA virus that reproduces by first transcribing its RNA into DNA then inserting the DNA molecule into a host DNA)
trans = across; **script** = write (transcription: the transfer of genetic information from DNA into an RNA molecule)

Key Terms

adenine (A)
AIDS
bacteriophages
cap
codon
cytosine (C)
DNA
DNA polymerase
double helix
emerging viruses
exons
genetic code
guanine (G)
HIV
introns
lysogenic cycle
lytic cycle
messenger RNA
molecular biology
mutagen
mutation
nucleotide
phages
polynucleotide
prion
promoter
prophage
provirus
retrovirus
reverse transcriptase
ribosomal RNA (rRNA)
RNA polymerase
RNA splicing
start codon
stop codon
sugar-phosphate backbone
tail
terminator
thymine (T)
transcription
transfer RNA (tRNA)
translation
uracil (U)
virus

Crossword Puzzle

Use the Key Terms list from this chapter to fill in the crossword puzzle.

ACROSS

4. the transcription enzyme used to link RNA nucleotides
5. the type of structural backbone of polynucleotides
7. the abbreviation for the type of polynucleotide that helps to form ribosomes
8. a nucleotide sequence that signals the end of a gene
9. the selective editing out of introns
10. viral DNA that inserts into a host genome
12. a monomer of a nucleic acid
13. a triplet that signals translation to start
15. the short name for viruses that attack bacteria
16. an infectious form of protein
18. viruses that attack bacteria
22. a physical or chemical agent that causes mutations
24. the abbreviation for the type of polynucleotide transcribed from a gene
25. a type of viral replication cycle that releases new phages by the death of the host cell
26. the nitrogenous base that pairs with guanine
27. a specific nucleotide sequence in DNA where transcription begins
29. portion of a gene that is excised from the RNA transcript
30. the set of rules relating nucleotide sequence to amino acid sequence
32. the nitrogenous base found only in RNA
33. the nitrogenous base that pairs with thymine
34. the enzyme that makes the covalent bonds during DNA replication
35. the three-base word used during translation
36. the nitrogenous base that pairs with cytosine
37. extra nucleotides added to the beginning of an RNA transcript
39. the type of virus that has appeared suddenly or recently comes to scientific attention
40. the nitrogenous base that pairs with adenine
41. the abbreviation for acquired immunodeficiency syndrome

DOWN

1. the transfer of genetic information from RNA into a protein
2. the abbreviation for the type of polynucleotide that serves as a molecular interpreter
3. the transfer of genetic information from DNA into an RNA molecule
4. an RNA virus that reproduces by mean of a DNA molecule
6. a triplet that signals translation to stop
10. phage DNA inserted into the bacterial chromosome
11. a change in the nucleotide sequence of DNA
14. the study of heredity at the molecular level
17. a polymer of nucleotides
19. an enzyme that catalyzes reverse transcription
20. the abbreviation for the human immunodeficiency virus
21. extra nucleotides added to the end of an RNA transcript
23. a type of viral replication cycle in which viral DNA replication occurs without phage production
28. the general shape of a DNA molecule
31. coding stretches of nucleotides in RNA transcripts
34. the abbreviation for the double-stranded nucleic acid stored in the nucleus
38. a bit of nucleic acid wrapped in a protein coat; these can cause serious disease

CHAPTER 11

How Genes Are Controlled

Studying Advice

a. The content of this chapter provides great depth and understanding to big questions in biology. For example: (1) What causes cancer, and *what can you do* to reduce your chances of getting cancer? (2) How do our cells specialize and arrange themselves as we develop? (3) How can we clone animals, and why would we want to do this? (4) What are stem cells, and how can they help treat disease? Although the material is challenging, it is a chance to understand the cutting edge of research in these important areas of biology and medicine.

b. Use the two following organizing tables to "keep track" of the details in the chapter. The tables can serve as important reference points as you read further and review for exams.

Student Media

Activities

The *lac* Operon in *E. coli*

Overview: Control of Gene Expression

Control of Transcription

Post-Transcriptional Control Mechanisms

Review: Control of Gene Expression

Signal-Transduction Pathways

Causes of Cancer

BLAST Animations

Signaling Across Membranes

Stem Cells

MP3 Tutors

Control of Gene Expression

Process of Science

How Do You Design a Gene Expression System?

Videos

Discovery Channel Video: Cloning

Discovery Channel Video: Fighting Cancer

C. elegans Crawling

C. elegans Embryo Development (time-lapse)

You Decide

Do Cell Phones Cause Brain Cancer?

Is Second-hand Smoke Dangerous?

Organizing Tables

TABLE 11.1 Describe the main functions of each component of the lactose (*lac*) operon and the regulatory genes and repressor proteins that affect the operon. Some of the cells of the table are already filled in to make this task a bit easier.

	Location	Function	What Controls Them
Promoter			
Operator			
Three enzyme genes	All three enzyme genes are located together next to the operator.		
Repressor proteins	The repressor proteins are free-floating in the cytoplasm.		Lactose, when present, binds and prevents the repressor from attaching to the promoter.

TABLE 11.2 Use the following table to compare the genes and process of gene regulation in prokaryotes and eukaryotes. Some of the cells of the table are already filled in to make this task a bit easier.

	Prokaryotes	**Eukaryotes**
Which group(s) use regulatory proteins that attach to DNA?		
Operons are typically used by which group(s)?		
Which are used more often, activators or repressors?		
Which group(s) modify mRNA before it is translated?	Prokaryotes do not usually modify mRNA.	
How long do the mRNA molecules last in the cell?	Prokaryotic mRNA is degraded by enzymes after only a few minutes.	

Content Quiz

Directions: Identify the *one* best answer for the multiple-choice questions. For true/false questions, determine if the statement is true or false. If false, change the underlined word(s) to make the statement true. Finally, add the correct word(s) to the fill-in-the-blank questions to make the statements true.

Biology and Society: Tobacco's Smoking Gun

1. Laboratory work demonstrated that a chemical in tobacco smoke causes ______________ that could lead to lung cancer.
   - A. mutations
   - B. viral infections
   - C. bacterial infections
   - D. overproduction of mucus
   - E. dehydration

2. True or False? A chemical found in cigarette smoke <u>has</u> been shown to specifically cause mutations that can lead to cancer.

3. Mutations in the p54 ______________ can lead to lung cancer.

How and Why Genes Are Regulated

PATTERNS OF GENE EXPRESSION IN DIFFERENTIATED CELLS

4. Genes determine the nucleotide sequences of:
   A. DNA.
   B. proteins.
   C. lipids.
   D. amino acids.
   E. mRNA.

5. Which of the following is the sequence of events involved in gene expression?
   A. DNA to RNA to proteins
   B. DNA to proteins to RNA
   C. RNA to DNA to proteins
   D. RNA to proteins to DNA
   E. proteins to RNA to DNA

6. True or False? The genes for specialized proteins are expressed in <u>all</u> cells.

7. Individual cells must undergo ______________; that is, they must become specialized in structure and function.

GENE REGULATION IN BACTERIA

8. In bacteria, gene expression is mainly controlled by:
   A. deleting certain genes from chromosomes.
   B. moving DNA into special capsules.
   C. limiting DNA replication.
   D. turning transcription on and off.
   E. making extra copies of chromosomes.

9. Which one of the following statements about the *lac* operon is *false*?
   A. Enzymes that help absorb and process lactose are produced by *E. coli* when lactose is absent.
   B. When RNA polymerase attaches to the promoter, it initiates transcription.
   C. The operator helps to determine whether RNA polymerase can attach to the promoter.
   D. The repressor protein binds to the operator and blocks the attachment of RNA polymerase to the promoter.
   E. When lactose is present, it interferes with the attachment of the *lac* repressor to the promoter by binding to the repressor and changing its shape.

10. The way that lactose works in the *lac* operon is most like:
    A. adding milk and sugar to coffee to improve its flavor.
    B. a boy distracting his mom while his brother takes some cookies.

C. cooking a meal and serving it to guests.

D. putting an ATM card into an ATM to get some money.

E. advertising a restaurant to attract customers.

11. True or False? RNA polymerase attaches to the <u>operator</u>.

12. A cluster of genes with related functions, along with the control sequences, is called a(n) ____________.

GENE REGULATION IN EUKARYOTIC CELLS

13. Examine Figure 11.3 in the text. A gene mutation would produce the greatest effects:

    A. when changes are made to the polypeptide in the cytoplasm.

    B. during translation.

    C. during processing of RNA in the nucleus.

    D. during transcription.

14. The extra X chromosome in human females is:

    A. expressed at about the same level as the other X chromosome.

    B. eliminated from the cell early in embryonic development.

    C. highly compacted and inactivated.

    D. less folded and more frequently expressed.

15. Eukaryotes usually:

    A. have operons.

    B. have a promoter and other control sequences for each gene.

    C. have regulatory fats that bind to DNA.

    D. have more repressor genes than activators.

16. The process of alternate RNA splicing in Figure 11.6 is most like:

    A. baking a pie and a cake and cutting them up to serve to guests.

    B. clipping news and sports articles out of newspapers to make two scrapbooks.

    C. sorting out beads to make a necklace.

    D. editing 8 hours of film in different ways to produce movies with different endings.

17. In eukaryotes, the most important stage for regulating gene expression is the:

    A. unpacking of chromosomal DNA.

    B. breakdown of mRNA.

    C. removal of introns from RNA.

    D. initiation of transcription.

    E. transport of mRNA from the nucleus to the cytoplasm.

18. Which one of the following is *not* a mechanism used to regulate gene expression after eukaryotic mRNA is transported to the cytoplasm?
    A. The mRNA molecule is typically broken down within hours to weeks.
    B. Different mRNA molecules combine in the cytoplasm to form new mRNA molecules.
    C. Proteins in the cytoplasm regulate translation.
    D. Proteins are edited after translation.
    E. Some final protein products last only a few minutes or hours.

19. True or False? DNA packing tends to <u>promote</u> gene expression.

20. True or False? Eukaryotic cells have <u>more</u> elaborate mechanisms than bacteria for regulating the expression of their genes.

21. True or False? Both prokaryotes and eukaryotes regulate transcription by using regulatory proteins that bind to <u>mRNA</u>.

22. Eukaryotic genes may be turned on when transcription factors bind to DNA sequences called ____________.

23. Repressor proteins may bind to DNA sequences called ____________, inhibiting the start of transcription.

CELL SIGNALING, HOMEOTIC GENES, AND DNA MICROARRAYS: VISUALIZING GENE EXPRESSION

24. Homeoboxes, found in homeotic genes:
    A. are types of proto-oncogenes that are a common cause of breast cancer.
    B. promote cancer by increasing the rate of cellular division.
    C. are characteristic of only mammals and birds.
    D. appear to be of recent evolutionary origin, coding for many animal traits that have only recently evolved.
    E. are very similar in many diverse organisms, suggesting a common evolutionary heritage.

25. When using a microarray, a researcher begins by collecting ____________ from a particular type of cell.
    A. mRNA
    B. DNA
    C. proteins
    D. lipids
    E. ATP

26. True or False? Homeoboxes containing homeotic genes have <u>different</u> developmental roles in mice and fruit flies.

27. True or False? Cell-to-cell signaling is a key mechanism in development and in the coordination of cellular activities throughout an organism's life.

28. True or False? A DNA microarray consists of a glass slide containing thousands of different single-stranded mRNA fragments arranged in a grid.

29. In cell-to-cell signaling, a signal molecule usually acts by binding to a receptor protein in the plasma membrane of the target cell and initiating a(n) ______________ pathway.

30. Researchers can use microarrays to learn what ______________ are active in different tissues.

31. A cell can secrete chemicals, such as ______________, that affect gene regulation in another cell.

Cloning Plants and Animals

THE GENETIC POTENTIAL OF CELLS, REPRODUCTIVE CLONING OF ANIMALS

32. In the process of nuclear transplantation, the nucleus from a donor cell is transplanted into:
    A. an egg in which the nucleus has been removed.
    B. a normal egg, allowing the two nuclei to fuse.
    C. the nucleus of another adult cell.
    D. another adult cell in which the nucleus has been removed.
    E. a sperm, which is used to fertilize an egg.

33. Reproductive cloning can be used to:
    A. produce herds of farm animals with desired traits.
    B. restock populations of endangered animals.
    C. produce pigs for organ donation that lack a gene that produces a protein that can cause immune system rejection in humans.
    D. All of the above.
    E. None of the above.

34. True or False? Hundreds or thousands of genetically identical clones can be produced from the cells of a single plant.

35. True or False? The process of cloning shows that differentiation does not involve irreversible changes in the DNA.

36. True or False? Dolly, the first mammal to be cloned, genetically resembled the egg donor.

37. Salamanders are capable of ______________, the regrowth of lost body parts.

38. True or False? Animal cloning was first performed in the 1990s.

THERAPEUTIC CLONING AND STEM CELLS

39. Which one of the following statements is *false*?
    A. When grown in laboratory culture, embryonic stem cells can divide indefinitely.
    B. Adult stem cells are much more difficult than embryonic stem cells to grow in culture.
    C. If the right conditions are used, scientists can induce changes in gene expression that cause differentiation of embryonic stem cells into a particular cell type.
    D. In the future, embryos may be created using a cell nucleus from a patient so that embryonic stem cells can be harvested and induced to develop into replacement tissues or organs.
    E. Embryonic stem cells are partway along the road to differentiation and usually give rise to only a few related types of specialized cells.

40. True or False? Adult <u>stem cells</u> generate replacements for nondividing differentiated cells in adults.

41. The purpose of ______________ cloning is to produce embryonic stem cells.

The Genetic Basis of Cancer

GENES THAT CAUSE CANCER

42. Which one of the following, if any, does *not* typically promote cancer?
    A. bacteria transmitting tumor-suppressor genes
    B. mutations in proto-oncogenes that code for growth factors
    C. the inactivation of tumor-suppressor genes that inhibit cellular growth
    D. All of the above typically promote cancer.

43. Cancers usually take a long time to develop because:
    A. oncogenes are rare.
    B. only old cells can become cancerous.
    C. several specific mutations must occur.
    D. cancer cells usually grow very slowly.

44. Finding a single cure for all cancer is unlikely because:
    A. we have made so little progress in recent years.
    B. we do not understand the genetics of the disease.
    C. the rapidly dividing cells cannot be killed.
    D. cancer is caused by many different factors.
    E. cancer cells migrate throughout the body.

45. True or False? Some viruses carry <u>oncogenes</u>.

46. True or False? For a proto-oncogene to become an oncogene, a cell's <u>RNA</u> must become mutated.

47. True or False? Cancer <u>always</u> results from changes in DNA.

48. True or False? Genetic screening of tumor cells allows for <u>personalized</u> therapies.

49. Many proto-oncogenes code for ____________, proteins that stimulate cell division.

CANCER RISK AND PREVENTION

50. The one substance known to cause more cases and types of cancer is:
    A. alcohol.
    B. asbestos.
    C. X-rays.
    D. table salt.
    E. tobacco.

51. Which one of the following is *not* a cancer risk factor?
    A. a diet high in fat
    B. a diet high in fiber
    C. the use of tobacco
    D. exposure to ultraviolet light
    E. alcohol consumption

52. True or False? People can reduce their risks of developing colon cancer by consuming a diet high in <u>sugar</u>.

53. Cancer-causing compounds called ____________ include ultraviolet light and tobacco smoke.

Evolution Connection: The Evolution of Cancer in the Body

54. Which of the following aspects of natural selection can also apply to cancer cells?
    A. the ability to produce more offspring than can be supported by the environment
    B. variation within the individuals of the population
    C. variations in the population must affect survival and reproductive success
    D. All of the above.
    E. None of the above.

55. True or False? Evolution drives the growth of <u>a tumor</u>.

56. Our understanding of cancer, like all other aspects of biology, benefits from a(n) ____________ perspective.

Word Roots

gen = producing (carcinogen: a cancer-causing agent)
onkos = tumor (oncogene: a gene that causes cancer)
proto = first (proto-oncogene: a normal gene with the potential to become an oncogene)

Key Terms

activators
adult stem cells
alternative RNA splicing
carcinogens
cellular differentiation
complementary DNA (cDNA)
DNA microarray
embryonic stem cells (ES cells)
enhancers
gene expression
gene regulation
growth factors
homeotic genes
nuclear transplantation
oncogene
operator
operon
promoter
proto-oncogene
regeneration
repressor
reproductive cloning
silencers
therapeutic cloning
transcription factors
tumor-suppressor genes
X chromosome inactivation

Crossword Puzzle

Use the Key Terms list from this chapter to fill in the crossword puzzle.

ACROSS

2. the type of gene that codes for proteins that normally help prevent uncontrolled cell growth
4. the replacement of body parts
6. the type of cloning that scientists use to help patients with irreversibly damaged tissues
7. DNA sequences where eukaryotic repressor proteins bind
8. using a somatic cell to make one or more genetically identical individuals
9. a type of switch between the promoter and the enzyme genes
11. a cluster of genes with related functions along with a promoter and an operator
12. DNA sequences where eukaryotic activators bind
14. the turning on and off genes
15. a glass slide containing thousands of kinds of single-stranded DNA fragments
17. a DNA molecule made in vitro using mRNA as a template and reverse transcriptase
18. cancer-causing agent
19. a type of transplantation technique used in animal cloning
20. the overall process by which genetic information flows from genes to proteins
21. the process by which one of two X chromosomes is inactivated at random
22. type of adult cell that generates replacements for nondividing differentiated cells
23. undifferentiated cells in an embryo that undergo unlimited division and produce several different types of cells
24. the process of cells becoming specialized in structure and function
25. a master control gene that regulates batteries of other genes
26. a protein secreted by certain cells that stimulates other cells to divide

DOWN

1. eukaryotic gene regulatory proteins
3. a normal gene with the potential to become an oncogene
5. a molecule that can turn off transcription
9. a cancer-causing gene
10. a site where the transcription enzyme RNA polymerase attaches and initiates transcription
13. a way for an organism to generate more than one polypeptide from a single gene
16. proteins that switch on a gene or group of genes

1
2
3
4
5
6
7
8
9
10
11
12
13
14
15
16
17
18
19
20
21
22
23
24
25
26

CHAPTER 12

DNA Technology

Studying Advice

a. This is a long and challenging chapter, but one with plenty of rewards. The medical, legal, and social issues related to genetic engineering are now a common part of our national dialogue. This chapter addresses issues straight out of today's headlines.

b. If you have not read and understood the major content in Chapters 10 and 11, review it before reading Chapter 12. Chapter 12 clearly builds on the information in these two prior chapters. The techniques described in Chapter 12 require a good understanding of molecular genetics.

c. This is not a chapter to study for the first time the night before an exam. If your instructor will address parts or all of this chapter in lecture, be sure to read this chapter *before* attending the lecture. Take notes on the techniques as you read. Make your own diagrams detailing the steps of these procedures and create your own mini-glossary to refer to as you read further and listen in lecture. This is one of the hottest fields in all of science, affecting what food we eat, how we fight disease, and the way we view ourselves.

Student Media

Activities

Applications of DNA Technology
Restriction Enzymes
Cloning a Gene in Bacteria
DNA Fingerprinting
Gel Electrophoresis of DNA
Analyzing DNA Fragments Using Gel Electrophoresis
The Human Genome Project: Human Chromosome 17
Making Decisions About DNA Technology: Golden Rice

BLAST Animations

Genetic Recombination in Bacteria
DNA Fingerprinting
Gel Electrophoresis

LabBench

Molecular Biology

MP3 Tutors

DNA Technology

Process of Science

How Can Antibiotic-Resistant Plasmids Transform *E. coli*?
How Can Gel Electrophoresis Be Used to Analyze DNA?

Videos

Discovery Channel Video: Transgenics
Discovery Channel Video: DNA Forensics
Biotechnology Lab

Organizing Tables

TABLE 12.1 Indicate the starting materials and products of each of the procedures described in the text. To help you in this task, the "procedure" cells in the table are already completed.

Process	Starting Materials	Procedure	Product
Cloning genes in a recombinant plasmid		1. Use the restriction enzyme to cleave the plasmid in only one place. 2. Use the same restriction enzyme to cleave the DNA into many pieces.	
Making pure genes using reverse transcriptase		1. Use reverse transcriptase to make a pure gene. 2. Insert the pure gene into a bacterium.	

TABLE 12.2 Describe the function of each of the enzymes listed in the table.

Enzyme	Enzyme Function
DNA ligase	
DNA polymerase	
Restriction enzymes	
Reverse transcriptase	

Content Quiz

Directions: Identify the *one* best answer for the multiple-choice questions. For true/false questions, determine if the statement is true or false. If false, change the underlined word(s) to make the statement true. Finally, add the correct word(s) to the fill-in-the-blank questions to make the statements true.

Biology and Society: DNA, Guilt, and Innocence

1. The Innocence Project is an organization that uses ______________ to exonerate prisoners.
   A. DNA profiling
   B. fingerprints
   C. lie detectors
   D. handwriting analysis

2. True or False? DNA profiling is the analysis of <u>protein</u> samples.

3. Using DNA technology, corn has been modified to produce its own ______________.

Recombinant DNA Technology

4. Recombinant DNA technology combines:
   A. genes from different sources.
   B. the nucleus of one cell with the cytoplasm of another.
   C. all of the genetic material of two cells.
   D. proteins from one cell and DNA from another.
   E. proteins from two different cells.

5. True or False? Biotechnology was first used thousands of years ago to make <u>vaccines</u>.

6. True or False? Recombinant DNA is formed when scientists combine pieces of <u>DNA</u> from two different sources.

7. Today, when people use the term *biotechnology,* they are usually referring to ______________, methods for studying and manipulating genetic material.

8. The direct manipulation of genes for practical purposes defines ______________.

APPLICATIONS: FROM HUMULIN TO FOODS TO "PHARM" ANIMALS

9. Which one of the following, if any, is *not* produced by recombinant DNA technology?
   A. humulin
   B. human growth hormone
   C. EPO
   D. saltwater-resistant plants
   E. All of the above are produced by recombinant DNA technology.

10. True or False? Today, <u>about half</u> of all American corn crops are genetically modified.

11. True or False? Genetically modified organisms have acquired one or more <u>cells</u> by artificial means.

12. Genetic engineering has produced rice that can help prevent ______________ deficiency, a disease that often leads to vision impairment.

13. Genetic engineering has now produced ______________ that synthesize and secrete humulin.

14. For many viral diseases, prevention by ______________ is the only medical way to fight the disease.

15. Organisms that have acquired one or more genes from another species by artificial means are called ______________ organisms.

16. In 2009, the FDA approved the production of an anticlotting protein that is produced by goats and is isolated from a goat's ______________.

RECOMBINANT DNA TECHNIQUES

Sequencing: Number the following seven steps in the order that they occur in the process of cloning genes in recombinant plasmids.

_____ 17. The recombinant DNA plasmids are mixed with bacteria. The bacteria take up the recombinant plasmids.

_____ 18. The plasmid and human DNA are cut.

_____ 19. The transgenic bacteria, with the desired gene, are grown in large tanks, producing large quantities of the desired protein.

_____ 20. A biologist isolates two kinds of DNA: many copies of a bacterial plasmid to serve as a vector and human DNA containing a gene of interest.

_____ 21. The bacterial clone with the specific gene of interest is identified.

_____ 22. The cut DNA is mixed. The human DNA and plasmids join together.

_____ 23. Each bacterium, with its recombinant plasmids, is allowed to reproduce.

24. When biologists want to customize bacteria to produce a specific protein, the gene for that protein is typically inserted into:
    A. the chromosome of another bacterium.
    B. the coat of a phage.
    C. the DNA of a phage.
    D. a plasmid.
    E. the chromosome of the bacterium.

25. True or False? The DNA of a plasmid is <u>part of</u> the bacterial chromosome.

26. True or False? A <u>nucleic acid probe</u> can be used to identify a bacterial clone carrying a particular gene of interest among the thousands of clones produced by shotgun cloning.

27. A(n) _______________ is a short, single-stranded molecule of radioactively labeled DNA whose nucleotide sequence is complementary to part of the gene or other DNA of interest.

28. The cutting tools for making recombinant DNA are bacterial enzymes called _______________ enzymes.

29. A collection of cloned DNA fragments that includes an organism's entire genome is called a(n) _______________.

30. When plasmids function as DNA carriers, moving genes from one cell to another, they are acting as _______________.

31. The workhorses of modern biotechnology are _______________.

32. A recombinant DNA molecule is produced when DNA pieces are connected into a continuous strand by _______________, which forms covalent bonds between adjacent nucleotides.

DNA Profiling and Forensic Science

INVESTIGATING MURDER, PATERNITY, AND ANCIENT DNA

33. Which one of the following, if any, is *not* a typical step in the DNA profiling process?
    A. DNA samples are collected from different sources.
    B. When the genetic material is insufficient to analyze, DNA in the sample is amplified.
    C. Proteins are produced from the DNA.
    D. The DNA samples are compared to each other.
    E. All of the above steps are used in a typical DNA profiling process.

34. True or False? To produce a DNA profile, scientists compare genetic markers, sequences in the genome that vary from person to person.

35. Since its introduction in 1986, _____________ has become a standard part of law enforcement.

DNA PROFILING TECHNIQUES

36. Which one of the following techniques is used to sort macromolecules, primarily on the basis of their electric charge and length?
   A. gel electrophoresis
   B. RFLP analysis
   C. recombinant DNA technology
   D. gene cloning
   E. polymerase chain reaction

37. The _____________ is a technique by which a specific segment of DNA can be targeted and copied quickly and precisely.

38. Repetitive DNA sequences from different individuals can be compared using _____________ analysis.

39. The key to PCR is an unusual _____________ that can withstand the heat needed to separate DNA strands.

40. Much of the DNA between genes is _____________ DNA, which can be used in STR analysis.

Genomics and Proteomics

THE HUMAN GENOME PROJECT

41. The first targets of genomics research were:
   A. algae.
   B. human cells.
   C. eukaryotic disease-causing microbes.
   D. prokaryotic disease-causing microbes.
   E. viruses.

42. Which one of the following is *false*?
   A. The human genome carries between 2,000 and 3,000 genes.
   B. The human genome contains approximately 3.2 billion nucleotide pairs.
   C. Much of the DNA of humans consists of introns and repetitive DNA.
   D. The genomes of many multicellular organisms have been sequenced.
   E. About 98% of the entire human genome consists of DNA that does not code for proteins.

43. True or False? The human genome consists of 24 different types of genes.

44. The science of studying whole genomes is called ____________.

45. Much of the DNA between genes consists of nucleotide sequences present in many copies, called ____________, sometimes used in DNA profiling.

TRACKING THE ANTHRAX KILLER

46. When the genomes of the anthrax spores used in the 2001 bioterrorist attacks were compared, it was determined that the mailed spores:
    A. were not identical.
    B. came from different sources.
    C. were all a harmless veterinary vaccine strain.
    D. all came from one particular flask.

47. True or False? Comparative genomics has revealed that humans and chimpanzees share 65% of their DNA.

GENOME-MAPPING TECHNIQUES

48. The goal of the Human ____________ Project is to collect information on all of the genetic variations that affect human health.
    A. Genome
    B. Variome
    C. Gefitinib
    D. Proteome
    E. Mapping

49. Genomes are most often sequenced using the:
    A. genetic mapping technique.
    B. DNA-sequencing method.
    C. physical mapping technique.
    D. whole-genome shotgun method.
    E. assembly method.

50. True or False? The functions of all human genes have been determined for the human genome.

51. True or False? The government-funded Human Genome Project used DNA from one human.

52. Research on lung cancer has revealed that the treatment of some types of cancer may be tailored to the specific ____________ of each patient.

PROTEOMICS

53. To better understand the structures and functions of cells and organisms, scientists are also studying proteomics because in organisms:
    A. proteins outnumber genes and proteins carry out cell activities.
    B. proteins outnumber genes and genes carry out cell activities.
    C. genes outnumber proteins and proteins carry out cell activities.
    D. genes outnumber proteins and genes carry out cell activities.

54. True or False? The study of genes and <u>the proteins they encode</u> is helping biologists understand how all of these parts interact within an organism.

55. The systematic study of full protein sets that a genome encodes is called ____________.

Human Gene Therapy

56. Which one of the following statements is *false?*
    A. The goal of gene therapy is to replace a mutant version of a gene with a properly functioning one within a living person.
    B. It was not until 2000 that the first scientifically strong evidence of effective gene therapy was reported.
    C. Safe and effective gene therapy is now widely used in medicine.
    D. Severe combined immunodeficiency (SCID) is a fatal inherited disease caused by a single defective gene.

57. True or False? A few of the SCID patients developed <u>cancer</u>, perhaps because the retrovirus used as a vector activated an oncogene.

58. One of the prime targets for gene therapy is bone ____________ cells that give rise to all the cells of the blood and immune system.

Safety and Ethical Issues

THE CONTROVERSY OVER GENETICALLY MODIFIED FOODS

59. Which one of the following statements is *false?*
    A. The European Union suspended the introduction of new GM crops.
    B. GM strains account for a significant percentage of several key agricultural crops.
    C. In the United States, labeling of GM foods is now being debated but has not yet become law.
    D. Lawn and crop grasses commonly exchange genes with wild relatives via pollen transfer.
    E. In the United States today, most public concern centers not on genetically modified (GM) foods but on recombinant microbes.

60. True or False? The U.S. National Academy of Sciences released a study finding <u>no scientific evidence</u> that transgenic crops pose any special health or environmental risks.

ETHICAL QUESTIONS RAISED BY DNA TECHNOLOGY

61. With respect to the ethical issues raised by genetic engineering, the authors argue that:
    A. there really is not much to be concerned about.
    B. the issues are so troubling that current research should be stopped.
    C. there are serious societal issues that need to be addressed.
    D. the scientific community will find answers to these concerns.
    E. scientists have not been concerned about these issues.

62. True or False? Genetic engineering of gametes (sperm or ova) has been accomplished in lab animals <u>but has not</u> been attempted in humans.

Evolution Connection: Profiling the Y Chromosome

63. Recent DNA profiles of the Y chromosome have revealed:
    A. the cause of many types of cancer.
    B. the common ancestry of thousands of men.
    C. the inability of the Y chromosome to direct protein production.
    D. surprising similarities between humans and mice.
    E. None of the above.

64. True or False? Barring mutations, the human Y chromosome passes <u>essentially intact</u> from father to son.

Word Roots

bio = life (biotechnology: the manipulation of living organisms to perform useful tasks)

liga = tied (DNA ligase: the enzyme that permanently "pastes" together DNA fragments)

trans = across; **genic** = producing (transgenic organism: an organism that contains genes from another organism)

Key Terms

biotechnology
clone
DNA ligase
DNA profiling
DNA technology
forensics
gel electrophoresis
gene cloning
genetic engineering
genetic marker
genetically modified (GM) organism
genomic library
genomics
human gene therapy
human genome project
nucleic acid probe
plasmid
polymerase chain reaction (PCR)
proteomics
recombinant DNA
repetitive DNA
restriction enzyme
restriction fragments
short tandem repeat (STR)
STR analysis
transgenic organism
vaccine
vector

Crossword Puzzle

Use the Key Terms list from this chapter to fill in the crossword puzzle.

ACROSS

2. the cutting tools for making recombinant DNA
4. a harmless variant or derivative of a pathogen used to prevent disease
6. much of the DNA between genes
7. another term for a genetically modified organism
8. a type of DNA molecule carrying genes derived from two or more sources
12. the study of whole sets of genes and their interactions
13. the study of full protein sets encoded by genomes
15. abbreviation for a technique that quickly and precisely copies a segment of DNA
16. the entire collection of cloned DNA fragments from a shotgun experiment
17. an international collaborative effort that sequenced the DNA of the entire human genome
18. a small ring of DNA separate from the chromosome(s)
19. methods for studying and manipulating genetic material
21. a specific pattern of electrophoresis bands that is of forensic use
23. the enzyme that permanently "pastes" together DNA fragments
24. the use of organisms to perform practical tasks
25. the scientific analysis of evidence for legal investigations
26. a method for sorting macromolecules based on their size and electric charge
27. the production of multiple copies of a gene

DOWN

1. abbreviation for the type of analysis that compares repetitive DNA sequences
3. in DNA technology, a labeled single-stranded nucleic acid molecule used to find a specific gene
5. the alternation of the genes of a person afflicted with a genetic disease
9. the role of a plasmid when it carries extra genes to another cell
10. any DNA segment that varies from person to person
11. a host that carries recombinant DNA
14. pieces of DNA produced by restriction enzymes
20. the direct manipulation of genes for practical purposes
22. a single organism that is genetically identical to another

CHAPTER 13

How Populations Evolve

Studying Advice

a. We all understand that we have ancestors. Our families extend back in time through our parents, grandparents, great grandparents, and their ancestors (although we soon lose track of their names). Why should the other species of the world be any different? Evolution is the study of the largest family of all, all of life on Earth.

b. A thorough understanding of evolution is necessary to make sense of the diversity of life on our planet. Just as we look to the history of a society to understand its customs, its language, and its form of government, we understand life with a sense of its past. This chapter lays the foundation for the common evolutionary theme of all the other chapters. It is largely conceptual with relatively few new terms. It is a chapter worth mastering.

c. The Hardy-Weinberg formula may at first seem intimidating. However, spend the time to understand how it can be used to make predictions about the chances of inheriting disease. Several questions are included here to quiz your comprehension.

Student Media

Activities

The Voyage of the *Beagle*: Darwin's Trip Around the World

Darwin and the Galápagos Islands

Reconstructing Forelimbs

Genetic Variation from Sexual Recombination

Causes of Evolutionary Change

BioFlix

Mechanisms of Evolution

Biology Labs On-Line

EvolutionLab
PopulationGeneticsLab

BLAST Animations

Evidence for Evolution: Homologous Limbs
Natural Selection
Evidence for Evolution: Antibiotic Resistance in Bacteria

LabBench

Population Genetics

MP3 Tutor

Natural Selection

Process of Science

How Do Environmental Changes Affect a Population?
What Are the Patterns of Antibiotic Resistance?
How Can Frequency of Alleles Be Calculated?

Videos

Discovery Channel Video: Charles Darwin
Discovery Channel Video: Antibiotics
Galápagos Islands Overview
Galápagos Marine Iguana
Galápagos Sea Lion
Galápagos Tortoise
Grand Canyon

Organizing Tables

TABLE 13.1 Define and compare the pairs of concepts in the table.

	Genetic Drift	Gene Flow
Mechanism: What is happening?		
Consequence: How does this affect the populations?		
	Founder Effect	**Bottleneck Effect**
Mechanism: What is happening?		
Consequence: How does this affect the populations?		
	Sexual Selection	**Natural Selection**
What is doing the selecting?		

TABLE 13.2 Compare the three types of selection in the table.

	Directional Selection	Disruptive Selection	Stabilizing Selection
What members of the population are being favored?			
How does the population curve change as a result of this selective force?			

Content Quiz

Directions: Identify the *one* best answer for the multiple-choice questions. For true/false questions, determine if the statement is true or false. If false, change the underlined word(s) to make the statement true. Finally, add the correct word(s) to the fill-in-the-blank questions to make the statements true.

Biology and Society: Persistent Pests

1. The World Health Organization's effort to spray mosquitoes' habitat with DDT in tropical regions of the world ended because:
   A. it was very successful and no longer needed.
   B. it killed the mosquitoes but also the animals that fed on them.
   C. the DDT accumulated in the ecosystem and caused too many other animals to die.
   D. it encouraged the evolution of new strains of mosquitoes resistant to DDT.

2. Pesticide resistance quickly evolves in insects because:
   A. the pesticide creates new individuals that can resist it.
   B. new species evolve that eat the pesticide.
   C. those that survive produce offspring that inherit genes for pesticide resistance.
   D. the pesticide changes the DNA and creates resistant individuals.
   E. the insects just get used to the presence of the insecticide.

3. True or False? After one application of a strong pesticide, the percentage of the population that is resistant to a pesticide will <u>decrease</u>.

4. True or False? Natural selection <u>creates</u> organisms that can resist pesticides.

Charles Darwin and *The Origin of Species*

DARWIN'S CULTURAL AND SCIENTIFIC CONTEXT, DESCENT WITH MODIFICATION

5. Which one of the following people helped set the stage for Darwin by proposing that adaptations evolve as a result of interactions between organisms and their environments?
   A. Buffon
   B. Lyell
   C. Lamarck
   D. Wallace
   E. Aristotle

6. In his book *The Origin of Species*, Charles Darwin developed two main points. These were that:
   A. the world is very old and evolution occurs by natural selection.
   B. evolution occurs slowly and new species form by spontaneous generation.
   C. new species evolve by mutations and new species do not reproduce with old species.
   D. modern species descended from ancestral species and organisms evolve by natural selection.

7. Which one of the following statements about Darwin's voyage on the *Beagle* is *false*?
   A. During the trip, Darwin read and was strongly influenced by the book *Principles of Geology* by Charles Lyell.
   B. The concept of natural selection came quickly to Darwin when he discovered 13 species of finches on the Galápagos.
   C. Darwin thought that most of the animals of the Galápagos resembled species living on the South American mainland.
   D. Darwin collected many South American species of plants and animals while on the trip.

8. In his book *Principles of Geology*, Lyell argued that:
   A. the Earth is about 6,000 years old.
   B. the Earth is ancient, sculpted by gradual geologic processes that continue today.
   C. fossils could be explained by a single worldwide flood that occurred in the last 10,000 years.
   D. erosion, earthquakes, and other geologic forces are new geologic events and did not occur in the past.

9. Which one of the following people also suggested the idea of natural selection?
   A. Buffon
   B. Lyell
   C. Lamarck
   D. Wallace
   E. Aristotle

10. True or False? Darwin <u>was</u> the first to suggest that life evolves.

11. True or False? In 1766, Buffon suggested that certain fossil forms might be <u>ancient versions of similar living species</u>.

12. True or False? <u>Lamarck</u> proposed that by using or not using its body parts, an individual develops certain characteristics, which it passes on to its offspring.

13. Lyell's principle of ______________ suggested that slow geological processes have sculpted the Earth.

14. Darwin proposed that all life is related by ______________ with modification.

Evidence of Evolution

THE FOSSIL RECORD

15. When we examine the fossil record, we find that:
    A. older fossils are in strata below younger fossils.
    B. birds appeared before dinosaurs.
    C. transitional fossil forms are missing.
    D. eukaryotes evolved before prokaryotes.
    E. there are major inconsistencies with molecular and cellular evidence.

16. True or False? The oldest known fossils are single-celled <u>eukaryotes</u>.

17. The oldest known fossils date to about ____________ billion years ago.

BIOGEOGRAPHY, COMPARATIVE ANATOMY

18. Which one of the following types of evidence first suggested to Darwin that modern organisms evolved from ancestral forms?
    A. comparative embryology
    B. the fossil record
    C. comparative anatomy
    D. biogeography
    E. molecular biology

19. Which one of the following is most like the concept of homology?
    A. mass producing a Toyota Prius
    B. repairing a car that has been in an accident
    C. different automakers producing their own versions of a pickup truck
    D. replacing worn tires, changing the oil, and tuning up the engine of a car

20. Which one of the following is most like the way that evolution works?
    A. remodeling your snow skis into water skis now that you've moved from Alaska to Florida
    B. building a house by gathering a group of architects and engineers to develop a new design
    C. mixing ingredients to bake a cake
    D. hiring a chemist to discover the secret formula of Coca-Cola

21. True or False? Biogeography <u>supports</u> the concept that species were individually placed into suitable environments.

COMPARATIVE EMBRYOLOGY, MOLECULAR BIOLOGY

22. The embryos of vertebrates:
    A. appear most similar at the earliest stages of development.
    B. appear most different at the earliest stages of development.
    C. are dramatically different, without any similar stages.

23. Evolution predicts that two closely related species, such as humans and chimpanzees, have:
    A. many similar genes and proteins.
    B. similar anatomical patterns.
    C. common embryological stages.
    D. a common fossil ancestor.
    E. All of the above.

24. True or False? Molecular biology confirmed the fossil record and other evidence supporting Darwin's view of the interrelatedness of all life.

25. Similar genes in two species suggest that the genes have been inherited from a(n) ____________.

Natural Selection

DARWIN'S THEORY OF NATURAL SELECTION

26. Which one of the following is *not* an assumption of natural selection?
    A. Populations typically produce more offspring than can survive.
    B. Resources to support a population are unlimited.
    C. Individuals in a population vary.
    D. Many traits are inherited.

27. Darwin suggested that the animals on the Galápagos Islands had become new species because they:
    A. were all much larger than their ancestors.
    B. were reproducing very quickly.
    C. ate foods similar to those eaten by their ancestors.
    D. were adapting to new, local environments.

28. True or False? The different beaks of the many Galápagos finch species are homologous, because they are variations of a common ancestral body plan.

29. True or False? The most fit individuals tend to leave the fewest fertile offspring.

30. Darwin thought that the Galápagos Islands were first colonized by animals from ____________.

31. According to natural selection, a population's inherent variability is screened by the ____________.

NATURAL SELECTION IN ACTION

32. Instead of being a creative mechanism, natural selection is really more a process of:
    A. editing.
    B. remixing.
    C. integrating.
    D. converging.
    E. directing.

33. True or False? An adaptation in one environment may be <u>useless or harmful</u> in different circumstances.

34. True or False? Doctors have documented an <u>increase</u> in drug-resistant strains of HIV.

EVOLUTIONARY TREES

35. In an evolutionary tree, typically turned sideways, the:
    A. oldest period of time is at the top.
    B. ancestors of groups are on the right.
    C. last common ancestor of all the groups is on the left.
    D. bottom group is always the oldest.
    E. None of the above is correct.

36. Evolutionary trees are often constructed based upon ______________ structures, using anatomical and molecular information.

37. Evolutionary trees represent ______________ relationships.

The Modern Synthesis: Darwinism Meets Genetics

POPULATIONS AS THE UNITS OF EVOLUTION

38. Which one of the following is the smallest unit of evolution?
    A. a cell
    B. an individual organism
    C. a population
    D. a species
    E. an ecosystem

39. True or False? Members of separate populations of the same species <u>cannot</u> interbreed.

40. True or False? <u>Individuals</u> evolve.

41. The total collection of alleles in a population at any one time is the ______________.

42. The fusion of genetics with evolutionary biology has become known as the ______________.

GENETIC VARIATION IN POPULATIONS

43. Which one of the following processes is *not* very significant in a large population in a single generation?
    A. mutation
    B. sexual recombination
    C. meiosis
    D. random fertilization

44. True or False? The genotype of an organism results from the combined influence of genetics and the environment.

45. Human height is an example of a(n) ____________ trait, which results from the combined effect of several genes.

ANALYZING GENE POOLS, POPULATION GENETICS AND HEALTH SCIENCE

Matching: Match each part of the Hardy-Weinberg formula, $p^2 + 2pq + q^2 = 1$, to what it represents.

_____ 46. p^2
_____ 47. $2pq$
_____ 48. q^2
_____ 49. 1

A. frequency of homozygous recessives
B. frequency of all genotypes in the gene pool
C. frequency of homozygous dominants
D. frequency of heterozygotes

Matching: Match the term or phrase on the left to its best description on the right.

_____ 50. population
_____ 51. population genetics
_____ 52. modern synthesis
_____ 53. gene pool
_____ 54. PKU

A. an inherited inability to break down the amino acid phenylalanine
B. the smallest biological unit that evolves
C. all the alleles in all individuals in a population
D. a field that studies genetic variation in a population
E. the combination of the work of Darwin and Mendel

55. If 1% of a population is homozygous recessive for a trait with two alleles, then:
    A. 63% are homozygous dominant.
    B. 49% are homozygous dominant.
    C. 18% are heterozygous.
    D. 25% are heterozygous.
    E. 81% are heterozygous.

MICROEVOLUTION AS CHANGE IN A GENE POOL

56. In a nonevolving population, from generation to generation, the frequency of alleles:
    A. changes, but the frequency of the genotypes stays the same.
    B. and the frequency of the genotypes stay the same.
    C. and the frequency of the genotypes change.
    D. stays the same, but the frequency of genotypes changes.

57. A nonevolving population in genetic equilibrium is also called the ______________ equilibrium.

58. A generation-to-generation change in a population's frequencies of alleles defines ______________.

Mechanisms of Evolution

GENETIC DRIFT, GENE FLOW

59. Imagine that a global disease spread quickly across Earth, killing every human except the people in your biology class. Although the future human population might someday recover, future generations would not represent the current diversity of all humans living today. This loss of diversity and change in the gene pool would be an example of:
    A. the founder effect.
    B. sexual recombination.
    C. gene flow.
    D. the bottleneck effect.
    E. Hardy-Weinberg equilibrium.

60. Your teacher takes your class on a spectacular field trip to study marine mammals, but your ship encounters a terrific storm. Your class and ship's crew land on a remote and uninhabited Pacific island where you live and prosper, never again making contact with other humans. Over many generations, the appearance of your island population takes on its own characteristics, representing the process of:
    A. the founder effect.
    B. sexual recombination.
    C. gene flow.
    D. the bottleneck effect.
    E. Hardy-Weinberg equilibrium

61. Which one of the following statements about gene flow is *false*?
    A. Modern transportation and the frequent relocation of people have increased gene flow in the human population.
    B. Gene flow occurs when fertile individuals move between populations.
    C. Populations may gain or lose alleles through gene flow.
    D. Gene flow tends to increase genetic differences between populations.

62. Which one of the following would *not* contribute to gene flow?

A. Very light seeds are easily blown across several miles in high storm winds.

B. A river floods over into nearby ponds.

C. A storm blows a flock of birds out to sea, where they land on an uninhabited island.

D. A giant redwood tree drops its seeds all around the base of its grand trunk.

63. True or False? Genetic bottlenecks usually <u>increase</u> the genetic variability in a population.

64. True or False? Some human diseases <u>are more abundant</u> in small populations due to genetic drift.

65. True or False? Genetic drift and gene flow <u>can contribute to</u> microevolution.

66. The founder effect and bottleneck effect are examples of ____________, an evolutionary mechanism by which the gene pool of a small population changes due to chance.

NATURAL SELECTION: A CLOSER LOOK

67. Which one of the following terms *best* represents the process of evolution?

A. deliberate

B. intentional

C. fixed

D. directional

E. responsive

68. Which one of the following promotes adaptation?

A. natural selection

B. mutations

C. the founder effect

D. gene flow

E. a genetic bottleneck

69. Which one of the following organisms has the greatest Darwinian fitness?

A. a robin that finds more food than any other female robin, but lays no eggs

B. a mother robin that has six chicks hatch, but only one that lives to reproduce

C. a mother robin that has two chicks, both of which live to reproduce

D. the largest mother robin in the entire population, who lays one large egg

E. a mother robin that lays eight eggs, but none of them hatch

Matching: Match the types of selection on the left to the descriptions on the right. One item on the left matches two items on the right.

_____ 70. directional selection

_____ 71. stabilizing selection

_____ 72. disruptive selection

A. selection favoring two or more contrasting morphs

B. selection favoring one extreme phenotype

C. selection favoring a particular trait within a narrow range

D. the most common type of selection

73. True or False? The Hardy-Weinberg equilibrium demands that all individuals in a population be equal in their ability to survive and reproduce.

74. Human birth weights are kept in the range of 3 to 4 kilograms by ____________ selection.

75. The fossil record indicates that when a population is challenged by a new environmental problem, the most common result is ____________.

SEXUAL SELECTION

76. A female bird preferring to mate with more brightly colored feathers on a male illustrates:
   A. gene flow.
   B. the bottleneck effect.
   C. the founder effect.
   D. sexual selection.
   E. genetic drift.

77. Distinct anatomical differences between males and females of a species are called sexual ____________.

Evolution Connection: The Genetics of the Sickle-Cell Allele

78. Which one of the following statements about sickle-cell disease is *false*?
   A. The sickle-cell allele is most common in Africa, where the malaria parasite is most common.
   B. People heterozygous for the sickle-cell allele are relatively resistant to malaria.
   C. People heterozygous for the sickle-cell allele are less common than people who are homozygous for the sickle-cell allele.
   D. People homozygous for the sickle-cell allele typically develop the disease.

79. The Hardy-Weinberg formula indicates that in a population with a sickle-cell allele frequency of 20%, ______________ of the population will be heterozygotes, benefiting from the sickle allele heterozygous protection, but at a cost of ______________ of the population getting sickle-cell disease.

   A. 32%; 4%
   B. 4%; 32%
   C. 40%; 10%
   D. 64%; 36%
   E. 36%; 64%

80. True or False? Heterozygous individuals are relatively resistant to malaria.

81. Sickle-cell disease is most common in tropical areas where ______________ is a major cause of death.

Word Roots

bio = life; **geo** = the earth (biogeography: the geographic distribution of species)
di = separate; **morph** = form (dimorphism: distinction in appearance based on secondary sexual characteristics)
homo = alike (homology: traits that appear similar due to common ancestry)
micro = small (microevolution: evolution at its smallest scale)
vestigi = **trace** (vestigial: historical remnants of structures that had important functions in ancestors)

Key Terms

biogeography
bottleneck effect
comparative anatomy
directional selection
disruptive selection
evolution
evolutionary adaptation
evolutionary tree
fitness
fossil record
fossils
founder effect
gene flow
gene pool
genetic drift
Hardy-Weinberg equilibrium
homology
microevolution
modern synthesis
natural selection
population
sexual dimorphism
sexual selection
stabilizing selection
vestigial structures

Crossword Puzzle

Use the Key Terms list from this chapter to fill in the crossword puzzle.

ACROSS

1. the comparison of body structures between different species
4. remnants of structures that served important functions in an organism's ancestors
5. genetic drift due to a drastic reduction in population size
6. the geographic distribution of species
9. distinction in appearance based on secondary sexual characteristics
13. the mechanism for descent with modification
14. a change in a gene pool of a small population due to chance
15. the type of selection that shifts the phenotypic curve of a population in favor of some extreme phenotype
16. the fusion of genetics with evolutionary biology
17. a form of natural selection in which individuals with certain characteristics are more likely than other individuals to obtain mates
18. a preserved imprint or remains of an organism that lived in the past
19. genetic drift in a new colony
20. genetic change in a population or species over generations
21. a branching diagram that reflects a hypothesis about evolutionary relationships among groups of organisms
22. a trait that appears similar due to common ancestry
23. a group of interacting individuals belonging to one species and living in the same geographic area
24. the ordered sequence of fossils as they appear in rock layers, marking the passage of geologic time

DOWN

2. a population's increase in the frequency of traits suited to the environment
3. the type of selection that maintains variation for a particular trait within a narrow range
7. all of the alleles in all of the individuals in a population
8. genetic exchange with another population
10. evolution at its smallest scale
11. the name of the formula and type of equilibrium that describes a population's gene pool
12. the type of selection that can lead to a balance of two or more different morphs
18. the contribution an individual makes to the gene pool of the next generation relative to the contributions of other individuals

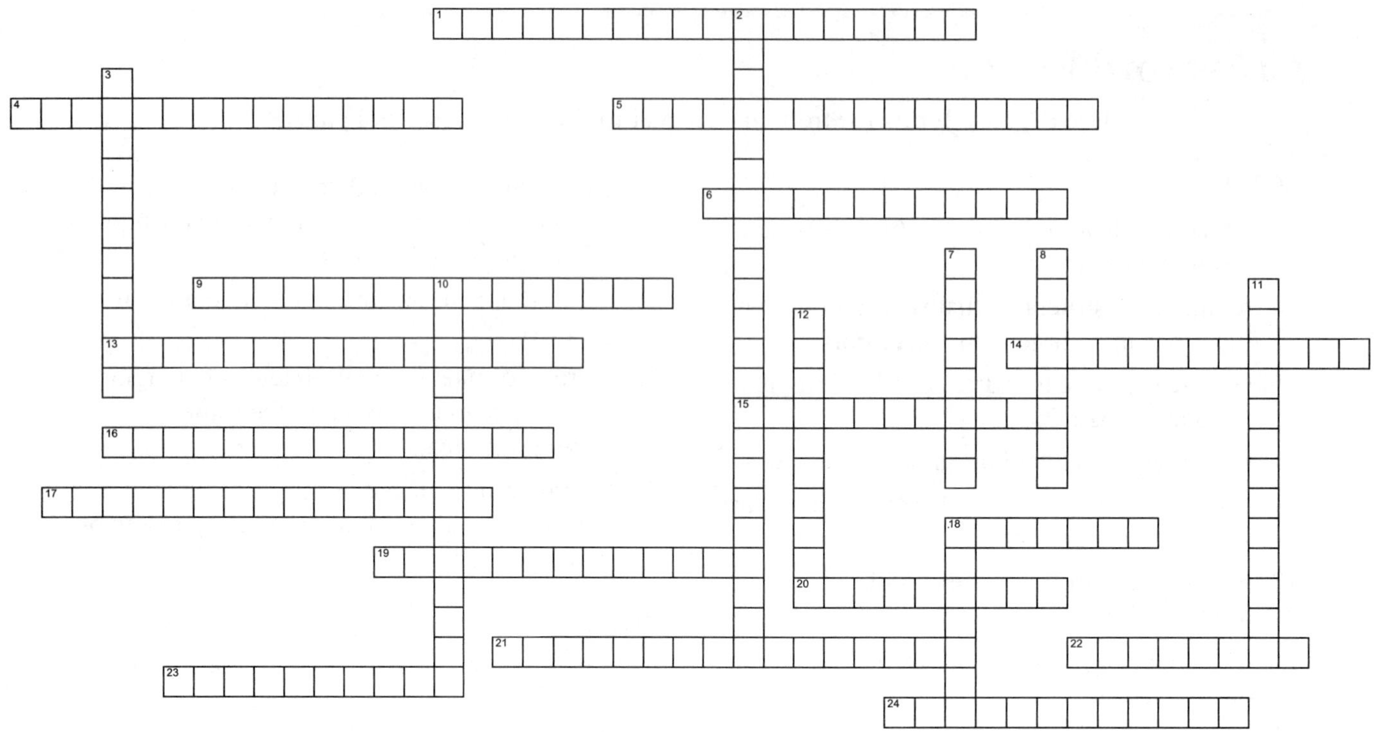
1
2
3
4
5
6
7
8
9
10
11
12
13
14
15
16
17
18
19
20
21
22
23
24

CHAPTER 14

How Biological Diversity Evolves

Studying Advice

a. Chapter 14 builds heavily upon Chapter 13. If, for some reason, you have not studied Chapter 13 before reaching this chapter, you should read Chapter 13 first.

b. Chapter 14 addresses some very broad and exciting questions about evolution: Why have major groups of organisms gone extinct? How do new species evolve? How have geological processes such as volcanoes, earthquakes, and continental drift affected the evolution of animals? Many students find this chapter to be one of the most interesting in the book. The new terminology is limited. So enjoy the ideas as you reflect on the major forces that have influenced the evolution of life on Earth!

Student Media

Activities

Overview of Macroevolution

Polyploid Plants

Allometric Growth

A Scrolling Geologic Record

Classification Schemes

Biology Labs On-Line

EvolutionLab

MP3 Tutors

Speciation

Process of Science

How Do New Species Arise by Genetic Isolation?

How Is Phylogeny Determined By Comparing Proteins?

Videos

Discovery Channel Video: Charles Darwin
Discovery Channel Video: Mass Extinction
Albatross Courtship Ritual
Blue-footed Boobies Courtship Ritual
Giraffe Courtship Ritual
Grand Canyon
Lava Flow
Volcanic Eruption
Galápagos Islands Overview

Organizing Tables

TABLE 14.1 List, with examples, the five prezygotic and three postzygotic barriers between species.

Prezygotic barriers	Examples
Postzygotic barriers	**Examples**

TABLE 14.2 Distinguish between allopatric and sympatric speciation, noting examples of each.

	Examples	Is it common in plants, animals, or both?
Allopatric speciation		
Sympatric speciation		

TABLE 14.3 Distinguish between the four main geologic divisions.

Geologic Divisions	Range of Time in Millions of Years Ago	Dominant Forms of Life
Cenozoic		
Mesozoic		
Paleozoic		
Precambrian		

Content Quiz

Directions: Identify the *one* best answer for the multiple-choice questions. For true/false questions, determine if the statement is true or false. If false, change the underlined word(s) to make the statement true. Finally, add the correct word(s) to the fill-in-the-blank questions to make the statements true.

Biology and Society: The Sixth Mass Extinction

1. Scientists think that we may be in the sixth greatest mass extinction in the last 600 million years mainly because:
   A. of the current rate of extinction.
   B. a mass extinction happens on a regular basis, and we are due for another.
   C. no new species seem to be forming.
   D. All of the above.
   E. None of the above.

2. True or False? Our current rate of extinction <u>has</u> reached the level of the other "big five" extinctions.

3. The fossil record reveals that ____________ is inevitable in a changing world.

Macroevolution and the Diversity of Life

4. Which one of the following, if any, is *not* included in the field of macroevolution?
   A. the multiplication of species
   B. the origin of evolutionary novelty
   C. the explosive diversification following some evolutionary breakthrough
   D. mass extinctions
   E. All of the above are included in the field of macroevolution.

5. True or False? Biological diversity is generated by <u>linear evolution</u>.

6. Darwin visited the volcanic ____________ Islands, named for their giant tortoises.

7. The major changes in the history of life, which include mass extinctions and the origin of evolutionary novelty, are collectively called ______________.

The Origin of Species

WHAT IS A SPECIES?

8. According to the "biological species concept," what keeps species separate?
   A. predators
   B. differences in diet
   C. natural selection
   D. reproductive barriers
   E. appearance

9. True or False? All humans belong to <u>the same</u> species.

10. True or False? Members of a biological species <u>cannot</u> interbreed with members of other species.

11. Because there can be no reproduction, biologists must distinguish ______________ species mainly by their differences in appearance.

REPRODUCTIVE BARRIERS BETWEEN SPECIES

12. Which one of the following is a postzygotic reproductive barrier?
   A. temporal isolation
   B. habitat isolation
   C. behavioral isolation
   D. hybrid inviability
   E. gamete isolation
   F. mechanical isolation

13. Two closely related species of tree frogs in Illinois cannot be distinguished from each other except for the male's chirping calls, used during the breeding season. This is an example of two species separated by:
   A. temporal isolation.
   B. habitat isolation.
   C. behavioral isolation.
   D. hybrid inviability.
   E. gamete isolation.
   F. mechanical isolation.

14. True or False? According to the biological species concept, the evolution of <u>reproductive barriers</u> is the key to the origin of new species.

15. A mule, formed by the hybridization of a horse and a male donkey, is an example of ______________, a form of reproductive isolation.

MECHANISMS OF SPECIATION

16. Sympatric speciation:
    - A. typically occurs over millions of years.
    - B. occurs when a population becomes geographically isolated.
    - C. is widespread among animals and rare in plants.
    - D. can occur in a single generation.
    - E. is the mechanism by which most Galápagos species have evolved.

17. In North America, some rare salamanders are triploid, possessing three copies of every chromosome. (They are an all-female species that reproduces by producing triploid eggs that develop without the addition of a sperm nucleus.) It is believed that these salamanders formed when the sperm from one species fertilized a diploid egg of another species. This evolutionary scenario is an example of:
    - A. hybrid sterility.
    - B. hybrid inviability.
    - C. allopatric speciation.
    - D. sympatric speciation.

18. True or False? Most polyploid species arise from <u>one</u> parent species.

19. Most of the plant species we grow for food are the result of ____________ speciation.

In the blanks for questions 20–24, write A if the example is of allopatric speciation or S if the example is of sympatric speciation.

_____ 20. A flock of birds is blown by a hurricane onto a remote Pacific island, where the birds remain isolated and become a new species.

_____ 21. A plant seed fails to undergo meiosis, fertilizing itself, and doubling the number of chromosomes in the next generation. These new plants do not interbreed with the parent species.

_____ 22. A long peninsula becomes a string of islands, separating several ant populations and evolving new ant species.

_____ 23. A glacier advances through the midwestern United States, splitting the populations of rabbits into eastern and western species.

_____ 24. Two species of moths hybridize, producing polyploid individuals that remain reproductively isolated and form a new polyploid species.

WHAT IS THE TEMPO OF SPECIATION?

25. The concept of punctuated equilibrium suggests that a species:
    - A. evolves quickly when it first forms, and then changes very little for a long time.
    - B. stays very similar when it first forms, but then undergoes drastic change, then is very similar just before becoming another species.
    - C. stays the same when it first evolves, then changes a lot and then not much at all, repeatedly, until it becomes another species.
    - D. changes a lot when it first evolves, then goes extinct.

26. The somewhat sudden switch in the music industry from vinyl records to compact discs is most like:
    - A. sympatric speciation.
    - B. allopatric speciation.
    - C. punctuated equilibrium.
    - D. microevolution.

27. True or False? A "sudden" geologic appearance of a species may actually be <u>50,000 to 100,000</u> years.

28. True or False? Once a species forms, species in a stable environment may experience <u>much</u> noticeable change.

The Evolution of Biological Novelty

ADAPTATION OF OLD STRUCTURES FOR NEW FUNCTIONS

29. Which one of the following terms is closest in meaning to the term <u>exaptation</u>?
    - A. building
    - B. remodeling
    - C. destroying
    - D. repairing
    - E. anticipating

30. True or False? Lightweight bones could <u>have</u> evolved in the ancestors of birds in anticipation of the evolution of flight in birds.

31. A structure that evolved in one context but conveyed advantages for other functions is a(n) ____________.

EVO-DEVO: DEVELOPMENT AND EVOLUTIONARY NOVELTY

32. Which one of the following is *not* a paedomorphic trait of humans?
    - A. large brain
    - B. rounded skull
    - C. large teeth
    - D. flat face
    - E. small jaws

33. The evolution of dramatically different species by the process of paedomorphosis is most like:
   A. ordering a birthday cake without frosting.
   B. blending a new flavor of wine by mixing wine from red and white grapes.
   C. adding extra luxuries to an automobile to make a special edition.
   D. repairing a truck after a collision.
   E. creating a tetraploid plant by hybridizing two diploid species.

34. Examine the grids in Figure 14.13. Which one of the following shows the *greatest* change in the development of the chimpanzee into an adult?
   A. top of the skull
   B. eye region
   C. region around the eyes
   D. region around the mouth (upper and lower jaw)
   E. bottom rear of the skull

35. True or False? Slight genetic changes in <u>homeotic genes</u> can dramatically change the appearance of an animal.

36. The retention of juvenile traits in the adult, or ____________, has occurred in salamanders and humans.

Earth History and Macroevolution

GEOLOGIC TIME AND THE FOSSIL RECORD

37. Pulling away many layers of wallpaper as a person remodels an old home is most like:
   A. the way that the continents formed.
   B. the distribution of fossils within a single sedimentary layer.
   C. exploring layers of sedimentary rock.
   D. the effects of earthquakes on the continents.
   E. radiometric dating.

38. Which one of the following commonly marks the boundaries of geologic eras?
   A. evidence of extensive earthquake activity
   B. sudden changes in ocean levels
   C. little change in biological diversity
   D. mass extinctions
   E. alignment of certain stars

39. Which one of the following is the correct sequence of the four major eras, from oldest to most recent?
   A. Paleozoic, Mesozoic, Precambrian, Cenozoic
   B. Precambrian, Paleozoic, Mesozoic, Cenozoic
   C. Cenozoic, Mesozoic, Paleozoic, Precambrian
   D. Cenozoic, Precambrian, Paleozoic, Mesozoic
   E. Precambrian, Mesozoic, Paleozoic, Cenozoic

40. Examine Figure 14.15 and read the figure legend to understand the half-life of an element. If an element has a half-life of 20,000 years, and we start out with 1 kilogram of the element, how much of the element will be present after 40,000 years?
   A. 4 kilograms
   B. 2 kilograms
   C. 0.5 kilograms
   D. 0.25 kilograms
   E. 0.10 kilograms

41. True or False? Examining the fossils in layers of sedimentary rock reveals the <u>absolute</u> age of the layers.

42. The rate of radioactive decay of specific isotopes can be used to date fossils in a process called ____________.

PLATE TECTONICS AND MACROEVOLUTION

43. Which one of the following did *not* occur as a result of the formation of Pangaea?
   A. Ocean levels were lowered.
   B. Shallow marine communities dried out.
   C. The total amount of shoreline was increased.
   D. Populations of organisms that had evolved in isolation were brought together.
   E. The continental interior increased in size.

44. If we compared the fossils of vertebrates found in Africa and South America, we would expect fossils found about 200 million years ago to be:
   A. similar, but those from animals 50 million years ago to be quite different.
   B. different, but those from animals 50 million years ago to be quite similar.
   C. similar, just like fossils from animals 50 million years ago.
   D. different, just like fossils from animals 50 million years ago.

45. True or False? The breakup of Pangaea probably caused <u>sympatric</u> speciation in many animals.

46. When Pangaea broke up, the continents moved apart because of ____________.

MASS EXTINCTIONS AND EXPLOSIVE DIVERSIFICATIONS OF LIFE

47. Which of the following is considered to be a contributing factor to the demise of the dinosaurs 65 million years ago?
   A. increased volcanic activity and continental movements
   B. a cooled global climate
   C. an asteroid or comet striking Earth
   D. All of the above are contributing factors.

48. Throughout the history of life on Earth, mass extinctions have been followed by periods of:

    A. little life.

    B. recovery of many of the same species.

    C. no life at all.

    D. an explosion of new diversity.

49. True or False? Over the last 600 million years, there have been <u>two</u> distinct periods of mass extinctions.

Classifying the Diversity of Life

SOME BASICS OF TAXONOMY

50. Which one of the following is the correct sequence of taxonomic groups moving from most general to most specific?

    A. domain, kingdom, phylum, class, order, family, genus, and species

    B. kingdom, domain, order, class, family, phylum, genus, and species

    C. phylum, kingdom, domain, class, order, genus, family, and species

    D. domain, kingdom, phylum, order, class, family, genus, and species

    E. class, kingdom, phylum, order, genus, family, species, and domain

51. Which one of the following is the correct way to write the scientific name for humans?

    A. *Homo Sapiens*

    B. homo Sapiens

    C. *homo sapiens*

    D. Homo sapiens

    E. *Homo sapiens*

52. The hierarchical system used to classify life is most like:

    A. the many ranks in the military.

    B. the many names for colleges in the United States.

    C. the different types of fruits and vegetables in a grocery store.

    D. the first, last, and middle names that people use.

53. True or False? The first part of the scientific name of several species in the same genus will <u>all be the same</u>.

54. Carolus Linnaeus suggested a system that gives each species a two-part name, or ______________.

CLASSIFICATION AND PHYLOGENY

55. Homologous structures are related to each other in the same way as:
    A. two different animal figures are to each other, when both are carved from bars of the same type of soap.
    B. two buildings are to each other, when one is made from brick and the other from steel.
    C. pens are to pencils.
    D. the brain is to the stomach.
    E. water is to fish.

56. Organisms that are very closely related are more likely to have:
    A. different genes and few homologous structures.
    B. different genes and many homologous structures.
    C. similar genes but few homologous structures.
    D. similar genes and many homologous structures.

57. Examine Figure 14.21 showing the classification of five carnivores. Which of the following pairs of genera are most closely related to each other?
    A. *Canis* and *Mephitis*
    B. *Canis* and *Lutra*
    C. *Panthera* and *Mephitis*
    D. *Panthera* and *Canis*
    E. *Mephitis* and *Lutra*

58. Examine the evolutionary relationships in Figure 14.23. The common ancestor of the group that included just the kangaroo and beaver had these characteristics:
    A. no hair, no mammary glands, and no gestation.
    B. hair, no mammary glands, and no gestation.
    C. hair, mammary glands, but no gestation.
    D. hair, mammary glands, and gestation.
    E. no hair, mammary glands, and egg laying.

59. True or False? Convergent evolution often produces structures that are homologous but not analogous.

60. Homologous structures have evolved from ____________ in a common ancestor.

61. Structures that are ____________ to each other are similar in function.

62. Cladistics is based upon ____________ relationships.

63. Very complex systems are less likely to be the result of ____________ evolution and more likely to be the result of ____________.

64. One of the most widely used methods in systematics, ____________, uses computers to analyze data to identify clades with unique homologies.

CLASSIFICATION: A WORK IN PROGRESS

65. In the three-domain system, the new domains Bacteria and Archaea are:
    A. subdivisions of the kingdom Protista.
    B. subdivisions of the kingdom Fungi.
    C. subdivisions of the kingdom Plantae.
    D. groups that were not discovered until the 1970s.
    E. groups that contain prokaryotic cells.

66. True or False? Although seaweeds are multicellular, they are commonly classified into the typically unicellular group Plantae.

67. The revision of the five-kingdom system to the three-domain system was largely based upon molecular studies and ____________.

Evolution Connection: Rise of the Mammals

68. Which of the following has most greatly impacted mammalian evolution?
    A. the breakup of Pangaea 180 million years ago
    B. increasing global temperatures
    C. the evolution of flowering plants
    D. the extinction of most dinosaurs
    E. None of the above.

69. True or False? In the past 50 million years, eutherian mammals have diversified much more than the egg-laying species of mammals.

70. Mass extinction leads to many vacant ecological ____________, which new spills fill as new forms of life evolve.

Word Roots

allo = other; **patric** = country (allopatric: the type of speciation in which a new species forms by geographic isolation)
bi = two; **nom** = name (binomial: a two-part name used to identify a species)
con = together (convergent evolution: the type of evolution in which unrelated organisms evolve structures with similar functions)
ex = beyond (exaptation: structures that evolve in one context and later become adapted for other functions)
macro = large (macroevolution: major biological changes evident in the fossil record)
paedo = child; **morphosis** = shaping (paedomorphosis: the retention in the adult of features that were juvenile in its ancestors)
post = after (postzygotic barriers: the type of barriers that prevent development of a zygote that is a hybrid between species)
pre = before (prezygotic barriers: the type of barriers that prevent the fertilization of the eggs of different species)
sym = together; **patri** = habitat (sympatric: the type of speciation in which a new species forms in the same geographic region)

Key Terms

allopatric speciation
analogy
binomial
biological species concept
clade
cladistics
class
convergent evolution
domain
evo-devo
family
genus
geologic time scale
kingdom
macroevolution
order
paedomorphosis
phylogenetic tree
phylum
postzygotic barriers
prezygotic barriers
punctuated equilibria
radiometric dating
reproductive barrier
speciation
species
sympatric speciation
systematics
taxonomy
three-domain system

Crossword Puzzle

Use the Key Terms list from this chapter to fill in the crossword puzzle.

ACROSS

2. a group of classes
6. the broadest category of classification that contains kingdoms
7. the system of classification that divides all life into three groups
8. a group of orders
9. a group of genera
11. a table of historical periods grouped into four eras
12. the type of systematics that uses computers and homologous structures
13. anatomical similarity due to convergent evolution
14. the study of the evolution of developmental processes in multicellular organisms
15. an ancestral species and all of its descendants
19. the type of evolution in which unrelated organisms evolve structures with similar functions
20. a two-part, Latinized name of a species; for example, *Homo sapiens*
23. a group of families
25. the type of barriers that prevent development of a zygote that is a hybrid between species
26. the process by which new species form in spurts of rapid change followed by periods of slow speciation
28. the study of the diversity and relationships of organisms, past and present
29. the identification, naming, and classification of species
30. a biological feature that prevents individuals of closely related species from interbreeding

DOWN

1. a branching diagram that represents a hypothesis about evolutionary relationships among organisms
3. a group of phyla
4. the type of barriers that prevent the fertilization of the egg of different species
5. the type of concept that defines a species
10. the type of speciation in which a new species forms by geographical isolation
16. major biological changes evident in the fossil record
17. the first part of a binomial
18. the retention of juvenile body features in the adult stage
21. in classification, the taxonomic category just below genus
22. the use of half-lives to identify the age of fossils
24. the origin of new species
27. the type of speciation in which a new species forms without geographical isolation

CHAPTER 15

The Evolution of Microbial Life

Studying Advice

a. Chapter 15 is the first of several survey chapters (Chapters 15, 16, and 17) that introduce quite a few names. Many students find that making note cards, with the group name on one side and its characteristics on the back, helps them organize the groups and learn them for exams. As you quiz yourself, remember to keep track of those questions you miss so that you can review those again. Consider alternating the way you quiz yourself, sometimes starting with the group name and other times starting with the group descriptions.

b. The five organizing tables that follow will help you review some of the key chapter information.

Student Media

Activities

The History of Life

Prokaryotic Cell Structure and Function

Classification of Prokaryotes

MP3 Tutors

Microbial Life

Process of Science

How Might Conditions on Early Earth Have Created Life?

What Are the Modes of Nutrition in Prokaryotes?

What Kinds of Protists Are Found in Various Habitats?

Videos

Discovery Channel Video: Early Life
Discovery Channel Video: Bacteria
Discovery Channel Video: Tasty Bacteria
Discovery Channel Video: Antibiotics
Cyanobacteria
Prokaryotic Flagella
Amoeba
Amoeba Pseudopodia
Paramecium Cilia
Paramecium Vacuole
Stentor
Stentor Ciliate Movement
Euglena Motion
Vorticella Cilia
Vorticella Detail
Vorticella Habitat
Plasmodial Slime Mold Streaming
Plasmodial Slime Mold
Dinoflagellate
Diatoms Moving
Various Diatoms
Water Mold Oogonium
Water Mold Zoospores
Chlamydomonas
Volvox Colony
Volvox Daughter
Volvox Flagella

Organizing Tables

TABLE 15.1 Identify the key events that occurred during the periods in Earth's history listed in the table.

Billions of Years Ago (BYA)	Key Events in Earth's History and the Evolution of Life
4.6	
4.0	
by 3.5	
3.5–2.5	
2.7	
2.1	
At least 1.2	
0.540 (540 million years ago)	
0.500 (500 million years ago)	

TABLE 15.2 Describe the four main stages, in order, of the hypothesis for the evolution of life on Earth.

	Key Events of Stage
Stage 1	
Stage 2	
Stage 3	
Stage 4	

TABLE 15.3 **Use this table to compare the prokaryotes and eukaryotes.**

	Prokaryotes	Eukaryotes
Domains		
Kingdoms	None indicated for prokaryotes.	
Examples of each kingdom		
Is a true nucleus present?	Yes or No	Yes or No
Are other membrane-enclosed organelles present?	Yes or No	Yes or No
Are cell walls present? If yes, are they made of cellulose?	Yes or No	Yes or No

TABLE 15.4 **Use this table to compare the different types of bacteria.**

Bacterial Cell Shape	Draw the Shape	Example from the Text
Cocci		
Staphylococci		
Streptococci		
Bacilli		No example given in the text.
Spirochetes		

TABLE 15.5 **Use this table to compare the types of nutritional strategies.**

Prokaryotic Group	Use Light Energy to Synthesize Organic Compounds?	Require at Least One Organic Nutrient as a Source of Carbon?	Examples Given in Text
Photoautotrophs	Yes or No	Yes or No	
Chemoautotrophs	Yes or No	Yes or No	
Photoheterotrophs	Yes or No	Yes or No	
Chemoheterotrophs	Yes or No	Yes or No	

Content Quiz

Directions: Identify the *one* best answer for the multiple-choice questions. For true/false questions, determine if the statement is true or false. If false, change the underlined word(s) to make the statement true. Finally, add the correct word(s) to the fill-in-the-blank questions to make the statements true.

Biology and Society: Can Life Be Created in the Lab?

1. What sort of artificial life are scientists trying to create?
   A. small, mouse-like critters that can eat garbage
   B. small, multicellular plants that can capture the sun's energy
   C. fungi that can help clean up toxic waste
   D. bacteria

2. True or False? Artificial life has been created in the laboratory.

3. Once scientists make an artificial genome, they hope to transplant it into a ____________ host cell.

Major Episodes in the History of Life

4. Which one of the following is the correct sequence of the evolution of early life?
   A. prokaryotic cells, photosynthetic prokaryotes, prokaryotes using cellular respiration, eukaryotic cells, multicellular eukaryotes
   B. photosynthetic prokaryotes, prokaryotic cells, eukaryotic cells, prokaryotes using cellular respiration, multicellular eukaryotes
   C. eukaryotic cells, photosynthetic eukaryotes, prokaryotes using cellular respiration, prokaryotic cells, multicellular prokaryotes
   D. multicellular eukaryotes, eukaryotic cells, prokaryotic cells, photosynthetic prokaryotes, prokaryotes using cellular respiration
   E. prokaryotic cells, prokaryotes using cellular respiration, photosynthetic prokaryotes, multicellular eukaryotes, eukaryotic cells

5. Which one of the following is *false*?
   A. Multicellularity evolved about 1.2 billion years ago.
   B. The oldest known eukaryotes are about 540 million years old.
   C. Life was confined to water for nearly 85% of its existence on Earth.
   D. Plants and fungi were the first to colonize land.
   E. Photosynthesis evolved about 2.7 billion years ago.

6. Examine the evolutionary relationships in Figure 15.1. Which one of the following pairs is most closely interrelated?
   A. archaea and plants
   B. archaea and protists
   C. bacteria and protists
   D. fungi and animals
   E. protists and animals

7. True or False? For almost 2 billion years, eukaryotes lived alone.

8. True or False? Mitochondria and chloroplasts are descendants of prokaryotes.

9. The addition of great amounts of oxygen into our atmosphere about 2.7 billion years ago favored organisms that could use the process of ____________.

The Origin of Life

RESOLVING THE BIOGENESIS PARADOX

10. Biogenesis is the idea that:
    A. life emerges by spontaneous generation.
    B. the first cells ever to evolve arose by spontaneous generation.
    C. it takes a male and female organism to produce new life.
    D. life is cellular.
    E. life gives rise to life.

11. True or False? The conditions on Earth when life first arose were very similar to conditions today.

12. The idea that life can come from nonliving matter, called ____________, was commonly accepted before modern science demonstrated otherwise.

A FOUR-STAGE HYPOTHESIS FOR THE ORIGIN OF LIFE, FROM CHEMICAL EVOLUTION TO DARWINIAN EVOLUTION

13. Which one of the following is the correct sequence of the four stages of the origin of the first cells?
    A. joining of monomers into polymers, abiotic synthesis of small organic molecules, packaging into pre-cells, and origin of self-replicating molecules
    B. origin of self-replicating molecules, joining of monomers into polymers, abiotic synthesis of small organic molecules, and packaging into pre-cells
    C. origin of self-replicating molecules, abiotic synthesis of small organic molecules, joining of monomers into polymers, and packaging into pre-cells
    D. abiotic synthesis of small organic molecules, joining of monomers into polymers, packaging into pre-cells, and origin of self-replicating molecules

14. Which of the following molecules were produced by experiments simulating conditions on primitive Earth?
    A. DNA and amino acids
    B. amino acids and sugars
    C. RNA and DNA
    D. cholesterol, DNA, and proteins

15. True or False? It appears that RNA evolved <u>before</u> DNA.

16. True or False? Natural selection <u>would not have</u> operated on the first pre-cells to have formed.

17. The early replication of RNA may have been catalyzed by ______________.

Prokaryotes

THEY'RE EVERYWHERE!

18. Which one of the following statements, if any, about prokaryotes is *false*?
    A. Some prokaryotes decompose organic matter and thus help recycle essential nutrients.
    B. Prokaryotes live in many places on Earth where eukaryotes cannot survive.
    C. Some prokaryotes cause serious disease.
    D. Some prokaryotes provide us with vitamins.
    E. All of the above are true.

19. True or False? Prokaryotes <u>could not</u> live today if all eukaryotes were eliminated.

20. True or False? Prokaryotes <u>outnumber</u> eukaryotes on Earth today.

21. Tuberculosis, cholera, food poisoning, and many sexually transmissible diseases are caused by ______________.

THE STRUCTURE AND FUNCTION OF PROKARYOTES

Matching: Match the word on the left to its best description on the right.

_____ 22. spirochete	A. chains of cocci bacteria
_____ 23. bacilli	B. clusters of cocci bacteria
_____ 24. cocci	C. rod-shaped bacteria
_____ 25. streptococci	D. dormant bacterial cells
_____ 26. staphylococci	E. spherical bacteria
_____ 27. endospores	F. spiral-shaped bacteria

Matching: Match the word on the left to its best description on the right.

_____ 28. photoheterotroph

_____ 29. chemoautotroph

_____ 30. chemoheterotroph

_____ 31. photoautotroph

A. energy source is sunlight; carbon source is CO_2

B. energy source is sunlight; carbon source is organic forms

C. energy source is inorganic compounds

D. energy source and carbon source are organic molecules

32. Which of the following nutritional strategies is (are) unique to prokaryotes?
    A. photoautotroph
    B. chemoheterotroph
    C. photoautotroph and chemoheterotroph
    D. photoheterotroph and chemoautotroph

33. Which one of the following statements about prokaryotes is *false?*
    A. Very few prokaryotic species are motile.
    B. Some prokaryotic species exhibit simple multicellular organization.
    C. Some endospores can remain dormant for centuries and survive boiling water.
    D. Many motile prokaryotes use flagella to move about.
    E. If resources are available, prokaryotic populations can double every 20 minutes.

34. Refrigeration retards food spoilage by:
    A. killing most bacteria on our foods.
    B. mutating bacteria with cold temperatures.
    C. reducing the rate of their reproduction.
    D. increasing the rate of binary fission.

35. True or False? Most bacteria have <u>a cell wall</u> exterior to their plasma membrane.

36. Prokaryotes reproduce by the process of _____________.

37. Most plants and algae are _____________.

38. All fungi and animals are _____________.

THE TWO MAIN BRANCHES OF PROKARYOTIC EVOLUTION: BACTERIA AND ARCHAEA

39. All bacteria and archaea:
    A. live in extreme environments.
    B. have the same structural and physiological characteristics.
    C. have a prokaryotic cellular organization.
    D. produce methane as a waste product of metabolism.
    E. aid in digestion in cattle.

40. True or False? Some archaea live in <u>aerobic</u> environments and give off methane as a by-product.

41. The ____________ are the prokaryotes most closely related to the eukaryotes.

BACTERIA AND HUMANS

42. Most pathogenic bacteria cause disease by:
   A. eating our living flesh.
   B. consuming so much oxygen that our tissues die.
   C. producing nitrogen-containing compounds.
   D. producing exotoxins and endotoxins.
   E. living in our cells and disrupting normal cellular processes.

43. Which one of the following is *not* a defense against bacterial disease?
   A. antibiotics
   B. endotoxins
   C. sanitation
   D. education

44. Which one of the following statements, if any, about bioterrorism is *false?*
   A. Early conquerors, settlers, and warring armies in South and North America gave native peoples items purposely contaminated with infectious bacteria.
   B. In the fall of 2001, five Americans died from the disease anthrax in a presumed terrorist attack.
   C. In 1984, members of a cult in Oregon contaminated restaurant salad bars with *Salmonella* bacteria.
   D. During the Middle Ages, armies hurled the bodies of plague victims into enemy ranks.
   E. All of the above are true.

45. True or False? Most bacteria <u>are</u> pathogenic.

46. Poisonous proteins secreted by bacterial cells are called ____________.

THE ECOLOGICAL IMPACT OF PROKARYOTES

47. Most of the nitrogen that plants use to make proteins and nucleic acids comes from:
   A. photoautotrophs that live in the air.
   B. prokaryotic metabolism in the soil.
   C. bioremediation.
   D. pathogenic bacteria that live inside leaves.
   E. genetically engineered bacteria that we add to the soil.

48. Bacteria are used to break down sewage and to clean up oil spills and other toxic environments through the process of ____________.

Protists

THE ORIGIN OF EUKARYOTIC CELLS

49. Eukaryotic cells evolved by:
    A. mitosis and meiosis.
    B. inward folds of the plasma membrane and endosymbiosis.
    C. inward folds of the plasma membrane and bioremediation.
    D. endosymbiosis and bioremediation.
    E. mitosis and endosymbiosis.

50. Which one of the following statements about chloroplasts and mitochondria is *false*? Both chloroplasts and mitochondria:
    A. possess DNA.
    B. make some of their own enzymes.
    C. reproduce by binary fission.
    D. appear to have evolved by endosymbiosis.
    E. are found in all eukaryotic cells.

51. Examine Figure 15.20. According to the endosymbiosis theory, the inner mitochondrial membrane corresponds to the:
    A. outer membrane of a chloroplast.
    B. nuclear membrane.
    C. original aerobic bacterial membrane.
    D. original membrane of a photosynthetic bacterium.
    E. endoplasmic reticulum.

52. Examine Figure 15.20 again. According to the endosymbiosis theory, the outer chloroplast membrane corresponds to the:
    A. inner mitochondrial membrane.
    B. nuclear membrane.
    C. plasma membrane of the cell that engulfed it.
    D. endoplasmic reticulum.
    E. outer membrane of an aerobic bacterium.

53. True or False? The ancestors of mitochondria were probably <u>anaerobic</u> bacteria.

54. Almost all eukaryotes have _____________, but only some have chloroplasts.

55. The first eukaryotic cells to evolve were _____________.

THE DIVERSITY OF PROTISTS

Matching: Match the group on the left to its best description on the right.

_____ 56. diatoms

_____ 57. flagellates

_____ 58. phytoplankton

_____ 59. dinoflagellates

_____ 60. amoebas

_____ 61. ciliates

_____ 62. plasmodial slime molds

_____ 63. apicomplexans

_____ 64. cellular slime molds

_____ 65. seaweeds

_____ 66. green algae

A. large, multicellular marine algae

B. protozoans covered by cilia

C. grass-green chloroplasts; includes *Volvox*

D. greatly flexible protozoans without permanent locomotory organelles

E. all parasitic protozoans with an apex for penetrating hosts

F. fungal life cycle with a plasmodium feeding stage

G. planktonic algae

H. silica cell wall that consists of two pieces

I. external plates of cellulose and two flagella

J. fungal life cycle with a solitary amoeboid feeding stage

K. protozoans that move with one or more flagella

Matching: Match the protist group on the left to its characteristic on the right.

_____ 67. flagellates

_____ 68. plasmodial slime molds

_____ 69. ciliates

_____ 70. seaweeds

_____ 71. apicomplexans

_____ 72. dinoflagellates

_____ 73. diatoms

A. mined and used as filtering material or an abrasive

B. their gel-forming substances are used in ice cream and pudding

C. responsible for red tides and massive fish kills

D. can be found among leaf litter on a forest floor

E. *Paramecium* is an example

F. includes the organism that causes malaria

G. includes *Giardia* and trypanosomes that cause sleeping sickness

74. Which of the following is (are) found in nearly all protists?
   A. multicellularity
   B. flagella
   C. chloroplasts
   D. mitochondria
   E. pseudopods
   F. cilia

75. Which one of the following groups consists of organisms that are decomposers?
   A. protozoans
   B. slime molds
   C. unicellular algae
   D. seaweeds

76. Which one of the following groups consists of multicellular photosynthetic protists?
   A. protozoans
   B. slime molds
   C. unicellular algae
   D. seaweeds

77. The cilia on the surface of ciliates move these protists in a way most similar to a:
   A. paddleboat.
   B. Viking ship with one hundred sets of oars.
   C. sailboat.
   D. tugboat pushing a ship out to sea.
   E. submarine using a nuclear engine to turn its propeller.

78. True or False? The closest relatives of seaweeds are <u>unicellular algae</u>.

79. The ______________ are photosynthetic protists that are most related to plants.

80. Seaweeds are classified into three groups based partly on the types of ______________ present in their chloroplasts.

Evolution Connection: The Origin of Multicellular Life

81. A major advantage of multicellularity is:
   A. the ability of cells to multiply.
   B. the smaller size of the organism.
   C. faster reproductive cycles.
   D. the ability for cells to specialize.

82. True or False? Multicellular organisms are fundamentally different from unicellular organisms.

83. The organism ______________ is a colonial green algae that shows some of the traits of early multicellular life.

Word Roots

api = the tip (apicomplexans: all parasitic protists named for an apparatus at their apex that is specialized for penetrating host cells)
archae = ancient (archaea: prokaryotic group most closely related to eukaryotes)
bacill = a little stick (bacilli: rod-shaped prokaryotes)
bi = two (binary fission: a form of asexual reproduction in which a single cell divides into two cells of about equal size)
bio = life; **genesis** = origin (biogenesis: the principle that life gives rise to life)
chemo = chemical; **auto** = self (chemoautotrophs: organisms that need only carbon dioxide as a carbon source and extract energy from inorganic substances)
dinos = whirling (dinoflagellates: protists with two flagella that cause them to spin)
endo = inner (endospores: bacterial resting cells)
exo = outside (exotoxins: toxic proteins secreted by bacterial cells)
flagell = a whip (flagellates: protozoa that move by means of one or more flagella)
hetero = different (chemoheterotrophs: organisms that must consume organic molecules for both energy and carbon)
patho = disease (pathogens: organisms that cause disease)
photo = light (photoautotrophs: photosynthetic organisms, including the cyanobacteria)
planktos = wandering (plankton: communities of mostly microscopic organisms that drift or swim near the water surface)
protos = first; **zoan** = animal (protozoan: early protists that ingested food)
pseudo = false; **pod** = foot (pseudopodia: temporary extensions of a cell used for locomotion and/or feeding)
spiro = spiral (spirochetes: spiral-shaped bacteria)
sym = together (endosymbiosis: the process whereby mitochondria and chloroplasts exist as associations between prokaryotic cells living within larger prokaryotic cells)

Key Terms

algae
amoebas
apicomplexans
archaea
bacilli
bacteria
binary fission
biogenesis
bioremediation
cellular slime molds
ciliates
cocci
diatoms
dinoflagellates
endospores
endosymbiosis
endotoxins
eukaryote
exotoxins
flagellates
green algae
pathogens
plankton
plasmodial slime mold
prokaryote
protists
protozoans
pseudopodia
ribozymes
seaweeds
spontaneous generation
symbiosis

Crossword Puzzle

Use the Key Terms list from this chapter to fill in the crossword puzzle.

ACROSS

1. the first eukaryote to evolve from prokaryotic ancestors
4. communities of mostly microscopic organisms that drift or swim near the water surface
5. type of protozoan that moves by means of cilia
6. algae and other organisms, mostly microscopic, that drift passively in ponds, lakes, and oceans
7. protozoans that move by means of one or more flagella
8. cells with true nuclei and many other organelles
9. large, multicellular marine alga
11. rod-shaped prokaryotic cells
13. a close association between organisms of two or more species
14. one of two prokaryotic domains, the other being the bacteria
17. the use of organisms to remove pollutants from water, air, and soil
21. life emerging from inanimate material
23. protist with a glassy silica-containing cell wall
24. protists, including Volvox, with grass-green chloroplasts
26. spherical species of bacteria
27. bacterial resting cell
28. the principle that life gives rise to life
29. toxic chemical component in the cell walls of certain bacteria
30. cellular extensions of amoeboid cells used in moving and feeding
31. protist that has a plasmodium feeding stage

DOWN

1. the first type of cell that lacks a true nucleus
2. disease-causing organism
3. organism with solitary amoeboid cells that can form a sluglike colony
10. an enzymatic RNA molecule that catalyzes reactions during RNA splicing
12. one of a group of parasitic protozoans, some of which cause human diseases
15. protist with two flagella
16. one of two prokaryotic domains, the other being the archaea
18. type of protist characterized by great flexibility and the presence of pseudopodia
19. type of cell division in which each daughter cell receives a copy of the single parental chromosome
20. photosynthetic, plantlike protist
22. the process that generated mitochondria and chloroplasts
25. toxic protein secreted by bacterial cells

1
2
3
4
5
6
7
8
9
10
11
12
13
14
15
16
17
18
19
20
21
22
23
24
25
26
27
28
29
30
31

CHAPTER 16

Plants, Fungi, and the Move onto Land

Studying Advice

a. The first three major chapter sections (through "Highlights of Plant Evolution") set the stage for the rest of the chapter. Read these first sections carefully, paying special attention to adaptations to life on land. The next four chapter sections ("Bryophytes" through "Angiosperms") survey these groups in the order they evolved. The adaptations described in the first part of the chapter are referred to regularly. The chapter ends with a brief survey of fungi.

b. If you have ever wondered why ferns, trees, mosses, and mushrooms look so different, this chapter will be interesting. Maybe you have suffered from athlete's foot, ringworm, or a yeast infection. Plants and fungi are all around us. This interesting chapter is a starting point for appreciating their diverse structures and roles in our world.

Student Media

Activities

Terrestrial Adaptations of Plants
Highlights of Plant Phylogeny
Moss Life Cycle
Fern Life Cycle
Pine Life Cycle
Angiosperm Life Cycle
Madagascar and the Biodiversity Crisis
Fungal Reproduction and Nutrition

Blast Animations

Alternation of Generations
Non-Flowering Plant Life Cycle
Flower Structure

Pollination and Fertilization
Flowering Plant Life Cycle

MP3 Tutors

Evolution of Plants
Alternation of Generations

Process of Science

What Are the Different Stages of a Fern Life Cycle?
How Are Trees Identified by Their Leaves?
How Does the Fungus *Pilobolus* Succeed as a Decomposer?

Videos

Discovery Channel Video: Plant Pollination
Discovery Channel Video: Trees
Discovery Channel Video: Colored Cotton
Discovery Channel Video: Fungi
Discovery Channel Video: Leafcutter Ants
Flower Blooming (time-lapse)
Bee Pollinating
Bat Pollinating Agave Plant
Flowering Plant Life Cycle (time-lapse)
Allomyces Zoospore Release
Phlyctochytrium Zoospore Release
Water Mold Oogonium
Water Mold Zoospores

Organizing Tables

TABLE 16.1 Describe the symbiotic relationships noted in the table.

Mutually beneficial symbiotic relationships	What organisms are involved?	How does each benefit from this relationship?
Mycorrhizae		
Plant pollination		
Lichens		
Endosymbiotic origin of mitochondria and chloroplasts (see Chapter 15)		

TABLE 16.2 Plant evolution: Indicate the living group that best represents each stage in the evolution of plants. Circle Yes or No to indicate the specific traits found in each group.

Four stages of plant evolution	List a living example	Gametangia?	Vascular tissue?	Are seeds present and enclosed in a special chamber?
First land plants		Yes or No	Yes or No	No seeds
Diversification of vascular plants with lignin		Yes or No	Yes or No	No seeds
Origin of seeds		Yes or No	Yes or No	Yes or No
Flowering plants		Yes or No	Yes or No	Yes or No

Content Quiz

Directions: Identify the *one* best answer for the multiple-choice questions. For true/false questions, determine if the statement is true or false. If false, change the underlined word(s) to make the statement true. Finally, add the correct word(s) to the fill-in-the-blank questions to make the statements true.

Biology and Society: Will the Blight End the Chestnut?

1. The American chestnut trees are seriously threatened because of:
   A. overharvesting by the timber industry during World War II.
   B. competition from Chinese elm trees introduced accidentally in the 1920s.
   C. an Asian fungus introduced into the United States about 1900.
   D. the overharvest of American chestnuts for food by early pioneers.
   E. their inability to tolerate global warming.

2. True or False? The American chestnut trees were destroyed by a <u>fungus</u>.

3. The chestnut blight is an example of ______________, a close association of two or more species.

Colonizing Land

TERRESTRIAL ADAPTATIONS OF PLANTS

4. The authors note that water lilies are like whales because both:
   A. need to come to the surface of the water to breathe.
   B. have very smooth surfaces.
   C. evolved from terrestrial ancestors.
   D. rely upon photosynthesis.
   E. are not found near the polar regions of Earth.

5. Plants are:
   A. unicellular prokaryotes that make organic molecules by aerobic respiration.
   B. unicellular eukaryotes that make organic molecules by photosynthesis.
   C. multicellular prokaryotes that make organic molecules by photosynthesis.
   D. multicellular eukaryotes that make organic molecules by aerobic respiration.
   E. multicellular eukaryotes that make organic molecules by photosynthesis.

6. Which of the following distinguishes plants from large algae?
   A. Plants use photosynthesis and algae do not.
   B. Plants are eukaryotes and algae are prokaryotes.
   C. Plants are multicellular and algae are unicellular.
   D. Plants have terrestrial adaptations and algae do not.
   E. Plants require oxygen and algae do not.

7. The fungus in mycorrhizae gives the plant:
   A. water and essential minerals, and in turn the fungus receives sugars from the plant.
   B. oxygen, and in turn the fungus receives water and sugars from the plant.
   C. carbon dioxide and carbon, and in turn the fungus receives sugars from the plant.
   D. sugars, and in turn the fungus receives water and essential minerals from the plant.
   E. sugars and essential minerals, and in turn the fungus receives water from the plant.

8. Vascular tissues in plants function most like:
   A. highways for distributing resources within a country.
   B. the Internet for quick and reliable communication between parts of our country.
   C. the steel infrastructure of a high-rise office building.
   D. the extensive air-duct system for moving air within a high-rise office building.
   E. an extensive sewage system draining the sewage systems of many apartments.

9. Lignin is to a plant cell as:
   A. muscle is to our bodies.
   B. calcium is to our bones.
   C. telephones are to society.
   D. blood is to our bodies.
   E. hemoglobin is to our blood.

10. The gametangia of plants function most like:
    A. the plastic wrapper around a loaf of bread.
    B. the circulatory system of a mouse.
    C. windows that open up in an apartment.
    D. a can opener.
    E. paddles in a canoe.

11. In most plants, fertilization:
    A. and development occur outside the female parent plant.
    B. occurs within, but development occurs outside the female parent plant.
    C. occurs outside, but development occurs within the female parent plant.
    D. and development occur within the female parent plant.

12. True or False? Mycorrhizae are found <u>on the roots</u> of some of the oldest plant fossils.

13. Most plants rely upon a symbiotic relationship between their roots and soil fungi, called ______________.

14. Carbon dioxide and oxygen move between the atmosphere and the interior of leaves via ______________, microscopic holes in the leaf surfaces.

15. Water loss is reduced from the leaf surface by the ______________, a waxy coating.

16. Plants use ______________ to gather resources below ground and ______________ to gather resources from above ground.

17. The chemical ______________ hardens plant cell walls, making them rigid.

THE ORIGIN OF PLANTS FROM GREEN ALGAE

18. The first terrestrial plants likely evolved from:
    A. algae that lived in shallow-water habitats.
    B. algae that lived in deep-water habitats.
    C. fungi that lived in shallow-water habitats.
    D. fungi that lived in deep-water habitats.
    E. prokaryotes that lived in deep-water habitats.

19. True or False? The first terrestrial plants faced <u>relatively few</u> pathogens and plant-eating animals.

20. The living green algae most closely related to plants are the ______________.

Plant Diversity

HIGHLIGHTS OF PLANT EVOLUTION

Matching: Match the group on the left to its best description on the right.

_____ 21. ferns
_____ 22. angiosperms
_____ 23. gymnosperms
_____ 24. bryophytes

A. early plants without true roots or leaves
B. vascular plants without seeds
C. vascular plants with "naked" seeds
D. vascular plants with flowers

25. True or False? Most plants alive today are <u>gymnosperms</u>.

26. An embryo packaged along with a store of food within a protective covering defines a(n) ______________.

BRYOPHYTES

27. Mosses display two key adaptations to life on land. These adaptations are:
    A. vascular tissue and lignin.
    B. vascular tissue and retention of embryos in the mother plant's gametangium.
    C. lignin and retention of embryos in the mother plant's gametangium.
    D. a waxy cuticle and retention of embryos in the mother plant's gametangium.
    E. a waxy cuticle and vascular tissue.

28. Examine Figure 16.10 showing the life cycle of a moss. Spores are produced by _____________ and sporophytes are produced by _____________.

A. mitosis; meiosis

B. meiosis; mitosis

C. mitosis; mitosis

D. meiosis; meiosis

29. True or False? The cells of the sporophyte are haploid.

30. True or False? Mosses and other bryophytes are unique in that the sporophyte is the dominant generation.

31. Spores are produced by the _____________, and gametes are produced by the _____________.

32. A life cycle in which the gametophyte and sporophyte stages produce each other is called _____________.

FERNS

33. Ferns are:

A. nonvascular, seedless plants with flagellated sperm.

B. vascular, seedless plants with flagellated sperm.

C. vascular, seedless plants that have pollen.

D. vascular plants with seeds.

E. vascular plants with seeds contained in special chambers.

34. True or False? The greatest amount of coal was deposited during the Carboniferous period.

35. Black sedimentary rock made up of fossilized plant material defines _____________.

GYMNOSPERMS

36. Which of the following types of climates favored the evolution of gymnosperms?

A. wetter and warmer

B. drier and warmer

C. drier and colder

D. wetter and colder

37. Which of the following adaptations to conserve water are found in gymnosperms?

A. broad leaves, no stomata, and a thick cuticle

B. broad leaves, stomata in pits, and a thin cuticle

C. thin leaves, stomata in pits, and a thin cuticle

D. thin leaves, stomata at the tips of bumps, and a thick cuticle

E. thin leaves, stomata in pits, and a thick cuticle

38. Compared to ferns, which one of the following was *not* a new terrestrial adaptation of gymnosperms?
   A. vascular tissue
   B. pollen
   C. further reduction of the gametophyte
   D. seeds

39. Examine the three variations on alternation of generation in plants in Figure 16.14. Which one of the figures best represents gymnosperms?
   A. A, a sporophyte dependent on a gametophyte
   B. B, a large sporophyte and a small, independent gametophyte
   C. C, a reduced gametophyte dependent on a sporophyte

40. True or False? A conifer is really a <u>sporophyte</u> with tiny <u>gametophytes</u> living in cones.

41. True or False? Conifers and other gymnosperms <u>have</u> ovaries.

42. Seeds develop from structures called ____________, located on the scales of female cones, in conifers.

43. The much-reduced gametophyte stage that houses cells that will develop into sperm is called ____________.

44. In gymnosperms, eggs develop within ____________.

ANGIOSPERMS

Matching: Match the flower part on the left to its best description on the right.

_____ 45. sepals
_____ 46. petals
_____ 47. stamen
_____ 48. anther
_____ 49. carpel
_____ 50. ovary

A. a protective chamber containing one or more ovules
B. a stalk bearing an anther
C. the style, with an ovary at the base and a stigma at its tip
D. usually green, they enclose the flower before it opens
E. the male organ in which pollen grains develop
F. usually the most attractive part of a flower, attracting pollinators

51. Which one of the following traits is found in angiosperms, but not gymnosperms?
   A. a seed enclosed within an ovary
   B. vascular tissue
   C. an adult gametophyte stage
   D. lignin supporting the cell walls
   E. stomata

52. What are the products of double fertilization?
    A. a zygote and endosperm
    B. fruit and endosperm
    C. a zygote and flower
    D. an embryo sac and pollen
    E. pollen and fruit

53. True or False? An angiosperm embryo is nourished by <u>endosperm</u>.

54. True or False? Most of our food comes from <u>gymnosperms</u>.

55. True or False? It is the <u>fruit</u> that accounts for the unparalleled success of the angiosperms.

56. The ripened ovary of a flower is the ____________.

57. Agriculture is a unique kind of evolutionary relationship between ____________ and animals.

PLANT DIVERSITY AS A NONRENEWABLE RESOURCE

58. Which one of the following is the leading cause of forest destruction?
    A. forest fires
    B. hurricanes and tornadoes
    C. human activities
    D. erosion
    E. pollution

59. True or False? At the current rate of destruction, Earth's <u>tropical rain forests</u> will be gone in 25 years.

60. True or False? Researchers have investigated the potential medical uses of <u>fewer than 2%</u> of the known plant species.

Fungi

CHARACTERISTICS OF FUNGI

61. Which one of the following statements about fungi is *false*?
    A. Fungi are mostly multicellular.
    B. Fungi are eukaryotes.
    C. Fungi are more closely related to plants than they are to animals.
    D. Fungi decompose organic matter and recycle the nutrients back to the soil.

62. Fungi reproduce by:
    A. producing spores by sexual or asexual reproduction.
    B. using flowers.
    C. producing seeds that are fertilized and develop within the female mushroom.
    D. special types of pollen that fertilize each other and then develop underground.
    E. budding or binary fission.

63. The relationship between a mushroom and its hyphae is most like the relationship between a:
    A. fish and a river.
    B. car and a highway.
    C. wristwatch and a person's arm.
    D. fire hydrant and the underground water pipes.
    E. book and a library.

64. True or False? The cell walls of fungi are made from <u>cellulose</u>.

65. The structures that account for wide dispersal of fungi are ____________.

66. Fungi feed by secreting powerful digestive ____________ into the environment and absorbing the digested compounds.

67. The bodies of most fungi are made from minute threads called ____________.

68. The feeding network of a fungus, composed of an interwoven mat of hyphae, is called a(n) ____________.

THE ECOLOGICAL IMPACT OF FUNGI

69. Which one of the following, if any, is *not* associated with fungi?
    A. mushroom
    B. production of the antibiotic penicillin
    C. athlete's foot
    D. ringworm
    E. vaginal yeast infection
    F. American chestnut blight
    G. truffle
    H. All of the above *are* associated with fungi.

70. True or False? Animals are <u>more</u> susceptible to parasitic fungi than are plants.

71. Along with bacteria and some invertebrate animals, fungi play an important ecological role as ____________ in natural environments.

72. One explanation for the unusual behavior of girls during the Salem witch hunt is that the girls had suffered from ____________ poisoning.

Evolution Connection: Mutually Beneficial Symbiosis

73. Which one of the following relationships is a type of mutually beneficial symbiosis?
    A. a thief stealing from a department store
    B. a person picking mushrooms to eat
    C. a hunter shooting ducks
    D. a person buying corn from a farmer
    E. a fungus growing on your shower curtain

74. True or False? A <u>parasitic</u> relationship is between organisms of different species that are in direct physical contact.

75. Mycorrhizae represent a mutually beneficial relationship between plant roots and ______________.

76. A(n) ______________ is composed of fungi and algae that live together in a mutually beneficial relationship.

Word Roots

angion = a container; **sperma** = seed (angiosperms: flowering plants)
bryo = moss; **phyte** = plant (bryophytes: a group of nonvascular plants that includes the mosses)
endo = inner; **sperm** = seed (endosperm: a source of nutrition for the developing embryo in a seed)
gamet = a wife or husband (gametangia: protective structures where plants produce their gametes)
gymno = naked (gymnosperms: vascular plants such as conifers with seeds not enclosed in specialized chambers)
myco = fungus; **rhiza** = root (mycorrhizae: symbiotic root–fungus combinations)
sporo = seed (sporophyte: the moss stage with diploid cells)
stoma = mouth (stomata: pores in leaves for gas exchange)
sym = together; **bio** = life (symbiosis: a relationship between two species in which one organism lives in or on another)

Key Terms

absorption	ferns	lignin	sepals
alternation of generations	filament	mosses	shoots
angiosperms	flower	mycelium	spores
anther	fossil fuels	mycorrhizae	sporophyte
bryophytes	fruit	ovary	stamen
carpel	fungi	ovules	stigma
charophytes	gametangia	petals	stomata
conifers	gametophyte	phloem	style
cuticle	germinate	plant	symbiosis
double fertilization	gymnosperms	pollen grain	vascular tissue
endosperm	hyphae	roots	xylem
	lichens	seed	

Crossword Puzzle

Use the Key Terms list from this chapter to fill in the crossword puzzle.

ACROSS

2. the densely branched network of hyphae in a fungus
3. ecological relationships between organisms of different species in direct contact
5. the female reproductive organ of a flower
7. energy deposits formed from the remains of extinct organisms
10. a moss stage with haploid cells
11. protective structures where plants produce their gametes
12. the portion of the vascular system in plants that transports sugar and other nutrients throughout the plant
13. one of a group of seedless vascular plants
15. a mature ovary of a flower that protects dormant seeds and aids in their dispersal
16. a short stem with four whorls of modified leaves
17. a chemical that hardens the cell walls of plants
19. a mechanism of fertilization in angiosperms, in which two sperm cells unite with two cells in the embryo sac to form the zygote and endosperm
20. minute threads of fungi that promote absorptive nutrition
21. a protective chamber containing one or more ovules
22. a mutualistic association between a plant root and fungus
23. vascular plants that bear naked seeds
25. the process by which small organic molecules are brought in from the surrounding medium
26. heterotrophic eukaryotes that acquire nutrients by absorption
27. a structure that develops in the plant ovary and contains the female gametophyte
28. the green algal group that is considered to be the closest relative of land plants
30. a moss, liverwort, or hornwort
31. the male organ in a flower in which pollen grains develop
33. a nutrient-rich tissue that provides nourishment to a developing embryo in angiosperm seeds
34. a multicellular eukaryote that makes organic molecules by photosynthesis
36. the moss stage with diploid cells
37. a plant embryo packaged along with a food supply within a protective coat

38. the tube-shaped, nonliving portion of the vascular system in plants that carries water and minerals from the roots to the rest of the plant
39. the stalk portion of the carpel
40. a plant structure that anchors a plant in soil, absorbs and transports minerals and water, and stores food
41. the sticky tip of a flower's carpel that traps pollen grains
42. a pollen-producing part of a flower that consists of a stalk and an anther
43. modified leaves, usually green, that enclose a flower before it opens

DOWN

1. a life cycle that switches between gametophyte and sporophyte stages
4. a haploid cell that divides mitotically to produce the gametophyte without fusing with another cell
5. cone-bearing plants
6. the aerial portion of a plant body, consisting of stems, leaves, and flowers
8. the stalk of a stamen
9. what a seed does when it first begins to grow
14. the group of vascular plants that bear naked seeds
18. the most familiar bryophyte group
24. a waxy coating of the leaves and other aerial parts of most plants
25. flowering plants
29. a system of tube-shaped cells that branches throughout the plant
32. typically the most striking parts of a flower, usually important in attracting insects
34. in a seed plant, the male gametophytes that develop within the anthers of stamens
35. microscopic pores in leaf surfaces that facilitate gas exchange
38. microscopic pores in leaf surfaces that facilitate gas exchange

1
2
3
4
5
6
7
8
9
10
11
12
13
14
15
16
17
18
19
20
21
22
23
24
25
26
27
28
29
30
31
32
33
34
35
36
37
38
39
40
41
42
43

CHAPTER 17

The Evolution of Animals

Studying Advice

a. Many students imagine biology as a chance to learn more about animals that they have seen, read about, watched on TV, or chased away when trying to enjoy time outdoors! If this sounds familiar, then you will enjoy this chapter, which introduces you to the amazing diversity of animal biology. At the end, you will also learn about human evolution and our place within the animal kingdom.

b. Take a big, deep breath before starting this, the longest chapter in the textbook. Chapter 17 introduces animal diversity with a survey of the major invertebrate and vertebrate groups. There is much to consider. Two pieces of advice:

   1. Do not try to study this chapter only a night or two before the exam. There are many names and characteristics to learn. You will need time to read, understand, and process this information. Students who wait too long to get started often confuse details on exams.
   2. Try to break this chapter down into smaller, more manageable "bites." Consider reading the sections one at a time, making note cards or other study tools as you read. Take short 5- to 10-minute breaks between sections.

c. The two following organizing tables will help you categorize the information in many chapter sections. Fill these tables in as you progress through the chapter.

Student Media

Activities

Animal Phylogenetic Tree

Characteristics of Invertebrates

Characteristics of Chordates

Primate Diversity

Human Evolution

MP3 Tutors

Human Evolution

Process of Science

How Are Insect Species Identified?

How Does Bone Structure Shed Light on the Origin of Birds?

Videos

Discovery Channel Video: Invertebrates

Hydra Budding

Hydra Eating *Daphnia* (time-lapse)

Hydra Releasing Sperm

Jelly Swimming

Thimble Jellies

Coral Reef

Nudibranchs

Earthworm Locomotion

C. elegans Crawling

C. elegans Embryo Development (time-lapse)

Tubeworms

Lobster Mouth Parts

Butterfly Emerging

Echinoderm Tube Feet

Manta Ray

Shark Eating a Seal

Clownfish and Anemone

Snake Ritual Wrestling

Galápagos Tortoise

Flapping Geese

Soaring Hawk

Swans Taking Flight

Bat Licking Nectar

Wolves Agonistic Behavior

Gibbons Brachiating

Chimp Agonistic Behavior

Chimp Cracking Nut

Organizing Tables

TABLE 17.1 Examine Figure 17.5 and the corresponding text pages to complete this table comparing invertebrate groups. Several cells are filled in to help you in this task.

Major Animal Group	Are True Tissues Present?	Type of Body Symmetry	Type of Digestive Tract	Examples of Animals in This Group
Sponges	Yes or No	No clear body symmetry	None	
Cnidarians	Yes or No			
Molluscs	Yes or No			
Flatworms	Yes or No		Gastrovascular cavity or no digestive tract at all	
Annelids	Yes or No			
Roundworms	Yes or No		Complete digestive tract with mouth and anus separate	
Arthropods	Yes or No			
Echinoderms	Yes or No			
Chordates	Yes or No			

TABLE 17.2 Examine Figure 17.29 and the corresponding text pages to complete the table comparing the vertebrate classes. One of the cells is already filled in to help you in this task.

Vertebrate Groups	Are Jaws Present?	Is the Skeleton Made of Bone or Cartilage?	Do the Adults Have Gills or Lungs?	Do They Have Legs? (If yes, how many do they walk on?)	Do They Have Amniotic Eggs?	Are They Ectothermic or Endothermic?
Jawless vertebrates						
Cartilaginous fishes						
Bony fishes				Most with gills, a few with lungs		
Amphibians						
Nonbird reptiles						
Birds						
Mammals						

Content Quiz

Directions: Identify the *one* best answer for the multiple-choice questions. For true/false questions, determine if the statement is true or false. If false, change the underlined word(s) to make the statement true. Finally, add the correct word(s) to the fill-in-the-blank questions to make the statements true.

Biology and Society: Rise of the Hobbit People

1. The text notes that the human fossils found on the Indonesian island of Flores might be:
   A. a new smaller species of humans or diseased members of our species.
   B. faked fossils that are most likely from a dog.
   C. faked fossils that are from a chimp or diseased members of our species.
   D. the fossils of children or another small ape.
   E. All of the above are false.

2. True or False? There are human disorders that can cause malformations similar to those found in the human bones discovered in Flores.

3. Biologists have discovered dwarf populations of deer, elephants, and hippos that are limited to living on ____________.

The Origins of Animal Diversity

WHAT IS AN ANIMAL?

4. Which one of the following sets of descriptions best describes animals?
   A. eukaryotic, unicellular, and heterotrophic organisms that use ingestion
   B. eukaryotic, multicellular, and autotrophic organisms that use egestion
   C. eukaryotic, multicellular, and autotrophic organisms that use ingestion
   D. eukaryotic, multicellular, and heterotrophic organisms that use ingestion
   E. prokaryotic, multicellular, and heterotrophic organisms that use ingestion

5. Examine Figure 17.2 showing the development of a sea star. In the early stages of development, the overall size of the embryo grows very little. Therefore, the average size of the embryonic cells:
   A. decreases.
   B. stays about the same.
   C. increases.

6. True or False? The life history of most animals includes a gastrula stage.

7. True or False? Animal development typically progresses from zygote to gastrula to blastula.

8. A change in body form, called ____________, remodels a larval stage into an adult form.

9. Most animals have ____________ cells for movement and ____________ cells that control them.

EARLY ANIMALS AND THE CAMBRIAN EXPLOSION

10. The Cambrian explosion:
    A. was a period of violent volcanic activity.
    B. resulted from Earth's impact with a large comet or asteroid, which killed off the dinosaurs.
    C. refers to a time when animal diversity increased dramatically.
    D. was a huge, internal blast deep within Earth that triggered tremendous earthquakes.
    E. resulted in the loss of most animal species and the origin of most plant species.

11. True or False? The cause of the Cambrian explosion is unknown.

12. True or False? Most zoologists now agree that the Cambrian organisms can be regarded as ancient representatives of <u>extinct</u> animal phyla.

ANIMAL PHYLOGENY

13. Which one of the following groups does *not* usually show bilateral symmetry?
    A. flatworms
    B. arthropods
    C. chordates
    D. annelids
    E. cnidarians

14. Which one of the following is *not* a function of a body cavity in at least some animals?
    A. cushions internal organs
    B. functions as a hydrostatic organ
    C. provides an extra space to store ingested food
    D. enables internal organs to move independently of the body surface

15. Which one of the following best demonstrates radial symmetry?
    A. an apple
    B. a skateboard
    C. a submarine sandwich
    D. a teapot
    E. a T-shirt

16. True or False? Most animals that move actively in their environment show <u>radial</u> symmetry.

17. A body cavity that is only partially lined by tissues derived from mesoderm is a(n) ____________.

Major Invertebrate Phyla

SPONGES

18. In sponges, food is distributed by:
    A. a primitive circulatory system with blood.
    B. a water vascular system.
    C. choanocytes.
    D. amoebocytes.
    E. tentacles.

19. Which one of the following is found in sponges?
   A. true tissues
   B. a coelom
   C. a digestive tract
   D. a skeleton
   E. gonads

20. True or False? Most sponges <u>live in fresh water</u>.

21. Sponge cells called _____________ filter food from the water.

CNIDARIANS

22. Which one of the following statements about cnidarians is *false?* Cnidarians:
   A. are carnivores.
   B. have a complete digestive tract with mouth and anus.
   C. occur in medusa and/or polyp form.
   D. have cnidocytes.
   E. are mostly marine.

23. Examine Figure 17.3. Which one of the following animals has a body plan that is most similar to the final stage of evolution indicated on the right side of this figure?
   A. a medusa of a jelly
   B. a shark
   C. an earthworm
   D. a grasshopper
   E. a sponge

24. True or False? The stinging cells of cnidarians are called <u>choanocytes</u>.

25. A sea anemone is an example of the _____________ body plan.

26. A jelly is an example of the _____________ body plan.

MOLLUSCS

27. The mollusc body consists of a:
   A. segmented tail, a muscular shield, and a mantle.
   B. muscular mass, a segmented foot, and a shield.
   C. muscular mass, a segmented foot, and a mantle.
   D. muscular foot, a visceral mass, and a mantle.
   E. muscular foot, a segmented visceral mass, and a shield.

28. True or False? Most <u>gastropods</u> have a shell divided into two halves hinged together.

29. The ______________ are a group of molluscs adapted for speed and agility.

30. Many molluscs use a(n) ______________, a strap-like rasping organ, to scrape up food.

31. Most molluscs have a shell, secreted by the ______________.

32. Molluscs house most of their internal organs in the ______________.

FLATWORMS

33. Flatworms are:

A. parasitic or free-living, bilaterally symmetrical, and unsegmented.
B. only free-living, bilaterally symmetrical, and segmented.
C. only free-living, radially symmetrical, and unsegmented.
D. only parasitic, bilaterally symmetrical, and segmented.
E. parasitic or free-living, radially symmetrical, and segmented.

34. Which of the following pairs of animals both have a gastrovascular cavity with a single opening?

A. sponges and planarians
B. hydras and tapeworms
C. jellies and planarians
D. tapeworms and sponges
E. sea anemones and sponges

35. True or False? Tapeworms do not have a gastrovascular cavity.

36. True or False? Tapeworms cannot infect humans.

37. Parasitic flatworms called ______________ are a major health problem in the tropics.

ANNELIDS

38. Which one of the following groups has segmental appendages that help the animals move and exchange gases?

A. earthworms
B. polychaetes
C. leeches
D. planarians
E. All of the above.

39. True or False? Most leeches are parasites.

40. True or False? A complete digestive tract allows food to move through the animal in only one direction.

41. Earthworms eat ______________ and eliminate ______________, a mixture of undigested material and mucus.

ROUNDWORMS

42. Which one of the following items is shaped most like a roundworm?
  A. a train with hundreds of railroad cars
  B. a round toothpick, tapered at both ends
  C. a spoon
  D. an apple
  E. a pancake

43. True or False? An acre of topsoil contains billions of nematodes.

44. Free-living roundworms in the soil are important ____________.

ARTHROPODS

45. Which one of the following is *not* a general characteristic of arthropods?
  A. segmentation
  B. jointed appendages
  C. exoskeleton
  D. six legs
  E. molting

46. Which one of the following habitats, if any, is *not* used by an arthropod?
  A. the surface of other animals
  B. the ocean
  C. underground
  D. the air
  E. All of these habitats are used by at least a few arthropods.

47. More than half of all known species of animals are:
  A. nematodes.
  B. insects.
  C. crustaceans.
  D. vertebrates.
  E. cnidarians.

Matching: Match the group on the left to its best description on the right.

_____ 48. insects
_____ 49. crustaceans
_____ 50. arachnids
_____ 51. millipedes and centipedes

A. adults with six legs and usually one or two pairs of wings
B. usually four pair of walking legs; includes spiders
C. multiple pairs of specialized appendages
D. similar segments along length of long body

52. True or False? Most animal species on Earth are arthropods.

53. Arthropods must molt their ____________ to grow.

54. The most dominant arthropods in the oceans are the ____________.

55. Arthropod ____________ and their appendages have become specialized for a great variety of functions.

56. Most arthropods belong to the group called ____________.

ECHINODERMS

57. Echinoderms are:

   A. unsegmented, marine animals with an endoskeleton and water vascular system.

   B. unsegmented, freshwater animals with an endoskeleton and circulatory system.

   C. segmented, freshwater and marine animals with an exoskeleton and circulatory system.

   D. segmented, freshwater and marine animals with an endoskeleton and water vascular system.

   E. segmented, marine animals with an exoskeleton and water vascular system.

58. True or False? Most echinoderms are sessile, or slow moving.

59. Most adult echinoderms have ____________ symmetry, but their larva typically show ____________ symmetry.

Matching: Match the invertebrate group on the left to its best description on the right.

_____ 60. sponges

_____ 61. jellies

_____ 62. corals

_____ 63. planarians

_____ 64. blood flukes

_____ 65. tapeworms

_____ 66. roundworms

_____ 67. gastropods

_____ 68. bivalves

_____ 69. cephalopods

_____ 70. leeches

A. mollusc group with a single, spiral shell

B. parasitic flatworms without a digestive tract

C. cylindrical body, tapered at both ends, and with a pseudocoelom

D. the group with the greatest number of species

E. segmented worm group that is mostly marine

F. a cnidarian group that is a polyp body form

Matching: Match the invertebrate group on the left to its best description on the right.

_____ 71. earthworms

_____ 72. polychaetes

_____ 73. arthropods

G. segmented worm group including bloodsuckers

H. mollusc group that includes squids and octopuses

I. segmented worm group that increases the fertility of soil

J. a cnidarian group that is a medusa body form

K. mollusc group with two shells hinged together

L. no true tissues; only freshwater or marine habitats

M. free-living flatworm group

N. parasitic flatworms that live inside blood vessels

Vertebrate Evolution and Diversity

CHARACTERISTICS OF CHORDATES

74. Which one of the following is *not* a chordate characteristic?
    A. ventral, solid nerve cord
    B. pharyngeal slits
    C. notochord
    D. post-anal tail

75. Examine the evolutionary relationships indicated in Figure 17.5. Which group is most closely related to the phylum Chordata?
    A. molluscs
    B. echinoderms
    C. annelids
    D. cnidarians
    E. arthropods

76. True or False? The phylum Chordata <u>does not include</u> invertebrates.

77. Other than vertebrates, the phylum Chordata includes _____________, which are long and thin, and _____________, which as adults are sessile filter feeders.

FISHES

Matching: Match the group on the left to its most distinct feature on the right.

_____ 78. ray-finned fishes

_____ 79. coelacanths

_____ 80. lungfishes

_____ 81. cartilaginous fishes

_____ 82. lampreys

_____ 83. hagfishes

A. no jaws; some are parasites attaching to large fish

B. jaws; swim bladder; deep-sea dwellers thought to be extinct

C. no jaws; use slime for defense

D. jaws; gulp air; live in Southern Hemisphere

E. jaws; no swim bladder; skeleton made of cartilage

F. jaws; swim bladder; greatest number of species

84. Which one of the following groups was the first to evolve?
    A. chondrichthyans
    B. ray-finned fishes
    C. lobe-finned fishes
    D. jawless vertebrates
    E. amphibians

85. True or False? If a shark stops swimming, it tends to <u>sink</u>.

86. Sharks sense changes in water pressure by using their ______________ system.

87. Bony fish have a protective covering over their gills called the ______________.

88. When not moving, bony fish do not sink because they have a(n) ______________.

AMPHIBIANS

89. Which one of the following characteristics is *not* lost during metamorphosis of a frog?
    A. tail
    B. gills
    C. legs
    D. lateral line system

90. During the life cycle of most frogs, the tadpoles:
    A. and adults are herbivores.
    B. and adults are carnivores.
    C. are carnivores and the adults are herbivores.
    D. are herbivores and the adults are carnivores.

91. Which one of the following groups is most likely to be the direct ancestor of amphibians?

A. chondrichthyans

B. ray-finned fishes

C. reptiles

D. jawless vertebrates

E. lobe-finned fishes

92. True or False? Animals with four limbs are called <u>hexapods</u>.

93. In addition to lungs, the ____________ of most adult amphibians promotes gas exchange.

REPTILES

94. Which of the following is a pair of reptilian adaptations for living on land?

A. amniotic eggs and dry scales

B. four legs and ears

C. lungs and a tail

D. lateral line system and ears

E. lateral line system and lungs

95. Examine the evolutionary relationships indicated in Figure 17.29. Within the vertebrates, which one of the following is a characteristic common and unique to reptiles, birds, and mammals?

A. vertebrae

B. hair

C. jaws

D. four legs

E. amniotic eggs

96. Which one of the following characteristics of birds is *not* found in other reptiles?

A. vertebrae

B. feathers

C. amniotic eggs

D. lungs

E. cranium

97. Which one of the following is *not* an adaptation for flight in birds?

A. honeycombed bones

B. a single ovary

C. loss of teeth

D. amniotic eggs

E. feathers

98. True or False? The most direct ancestors of birds were small, <u>two-legged dinosaurs</u>.

99. True or False? Because they are ectotherms, reptiles require <u>more</u> calories than a mammal of similar size.

100. Because of their dry skin, reptiles get most of their oxygen through their _____________.

101. The evolution of the _____________ egg allowed reptiles to reproduce without returning to water.

102. Birds are _____________, using their own metabolic heat to maintain a warm, steady body temperature.

MAMMALS

103. Mammals are primarily:
   A. terrestrial and endothermic.
   B. terrestrial and ectothermic.
   C. aquatic and endothermic.
   D. aquatic and ectothermic.

104. Which one of the following is *not* a characteristic of all mammals?
   A. hair
   B. mammary glands
   C. endothermy
   D. a placenta

105. True or False? A brief gestation followed by development in a pouch is characteristic of <u>eutherian</u> mammals.

106. The mammals that lay eggs are the _____________.

The Human Ancestry

THE EVOLUTION OF PRIMATES

107. Which one of the following is *not* a primate adaptation for living in trees?
   A. rigid shoulder joints
   B. eyes close together in the front of the face
   C. excellent hand-eye coordination
   D. extensive parental care
   E. agile hands

108. Which one of the following is a characteristic of Old World but *not* New World monkeys?

   A. ground-dwelling
   B. prehensile tails
   C. single births with a long period of nurturing
   D. nails instead of claws

109. True or False? In general, apes <u>are larger than</u> monkeys.

110. True or False? Modern apes live only in tropical regions of the <u>New World</u>.

111. Humans, apes, and monkeys form the group called ____________.

112. Of all the apes, only gibbons and orangutans are primarily ____________.

THE EMERGENCE OF HUMANKIND, THE PROCESS OF SCIENCE: WHAT DID NEANDERTHALS LOOK LIKE?

Matching: Match the group on the left to its best description on the right.

_____ 113. *Homo habilis*

_____ 114. *Homo neanderthalensis*

_____ 115. *Homo erectus*

_____ 116. *Australopithecus afarensis*

_____ 117. *Homo sapiens*

A. the species of Lucy, about 3.2 million years old

B. the species of modern humans

C. about 2.4 million years old, an early tool user

D. the first species to extend humanity's range beyond Africa

E. large-brained, skilled toolmakers, living 200,000 to 35,000 years ago

118. *Homo sapiens* appears to have evolved:

   A. about 195,000 years ago in what is now modern Ethiopia.
   B. 500,000 years ago from *Homo neanderthalensis* in what is now modern Europe.
   C. 1.8 million years ago from *Homo erectus* in what is now modern Asia.
   D. 2.4 million years ago from *Australopithecus afarensis* in East Africa.
   E. about 3.2 million years ago in southern Africa.

119. The pattern of the history of human evolution is most like:

   A. a ladder.
   B. a marching band in a parade.
   C. a bush.
   D. people lined up to go through a door.

120. Which one of the following is the correct sequence in the evolution of human culture?
  A. agriculture, hunting-and-gathering nomads, Industrial Revolution
  B. Industrial Revolution, agriculture, hunting-and-gathering nomads
  C. hunting-and-gathering nomads, Industrial Revolution, agriculture
  D. agriculture, Industrial Revolution, hunting-and-gathering nomads
  E. hunting-and-gathering nomads, agriculture, Industrial Revolution

121. Recent genetic analysis of Neanderthal DNA indicates that they:
  A. had red hair and pale skin.
  B. had arms that were much longer than their legs.
  C. had little hair on their bodies.
  D. were not able to walk upright.
  E. were not able to see very well.

122. True or False? Chimps <u>are</u> the parent species of humans.

123. True or False? Most of the distinct human traits evolved <u>at the same time</u>.

124. True or False? Upright posture evolved in humans <u>before</u> our enlarged brain.

125. True or False? The brains of Neanderthals were <u>about the same size as</u> the brains of modern humans.

126. Fossil evidence indicates that bipedalism in humans evolved at least ______________ million years ago.

127. Language, written and spoken, is the main way that ______________ is transmitted.

Evolution Connection: Recent Human Evolution

128. The FOXP2 gene is most related to the ability to:
  A. walk.
  B. read and write.
  C. speak.
  D. see well.
  E. make tools.

129. True or False? The FOXP2 gene <u>was</u> found in the DNA of Neanderthals.

Word Roots

amphi = double; **bio** = life (amphibians: vertebrates that live life in and out of the water)

annel = ring (annelids: segmented worms)

anthrop = man; **oid** = like (anthropoid: monkeys, apes, and humans)

arachn = spider (arachnids: spiders, scorpions, mites, and ticks)

arthro = joint; **pod** = foot (Arthropoda: the phylum of animals with jointed legs)

bi = double; **valva** = leaf of a folding door (bivalves: molluscs with two hinged shells)

centi = one hundred; **ped** = foot (centipedes: arthropods with many legs, one pair per body segment)

ceph(al) = head (cephalopods: molluscs that include octopus and squid)

echino = spiny; **derm** = skin (Echinodermata: the phylum of sea stars, which have spiny skin)

ecto = outside; **therm** = temperature (ectotherms: animals that absorb heat instead of producing it internally)

endo = inside (endothermic: animals that produce heat internally)

exo = outside (exoskeleton: an external skeleton such as that found in all arthropods)

gaster = belly (gastropods: molluscs with their gut near their muscular foot)

homin = man (hominids: humans and our nearest ancestors)

marsupi = bag or pouch (marsupials: the pouched mammals)

meta = change; **morph** = form (metamorphosis: a change in animal form during a lifetime)

milli = one thousand (millipedes: arthropods with many legs, two pairs per body segment)

mollusc = soft (Mollusca: the phylum of soft-bodied animals usually surrounded by hard shells)

mono = one (monotremes: mammals that lay eggs)

noto = the back; **chord** = string (notochord: a flexible, longitudinal rod characteristic of chordate animals)

platy = flat; **helminthes** = a worm (Platyhelminthes: the phylum of flatworms)

por = pore; **fer** = bearer (Porifera: the phylum of sponges)

post = behind (post-anal tail: the type of tail found in chordate animals)

pseudo = false (pseudocoelom: a body cavity that is not completely lined by tissue derived from mesoderm)

tetra = four (tetrapods: the group of terrestrial vertebrates with four legs)

Key Terms

amniotes
amniotic egg
amphibians
animals
annelids
anthropoids
arachnids
arthropods
bilateral symmetry
birds
bivalves
blastula
body cavity
body segmentation
bony fishes
cartilaginous fishes
centipedes
cephalopods
chordates
cnidarians
coelom
complete digestive tract
crustaceans
culture
dorsal, hollow nerve cord
earthworms
echinoderms
ectotherms
endoskeleton
endotherms
eutherians
evo-devo
exoskeleton
flatworms
gastropods
gastrovascular cavity
gastrula
hominids
hominoids
insects
invertebrates
larva
lateral line system
leeches
lobe-finned fishes
lungfishes
mammals
mantle
marsupials
medusa
metamorphosis
millipedes
molluscs
monotremes
nematodes
notochord
operculum
pharyngeal slits
placenta
placental mammals
polychaetes
polyp
post-anal tail
primates
pseudocoelom
radial symmetry
radula
ray-finned fishes
reptiles
roundworms
sponges
swim bladder
tetrapods
vertebrates
water vascular system

Crossword Puzzle

Use the Key Terms list from this chapter to fill in the crossword puzzle.

ACROSS

1. an internal skeleton
3. a group of annelids notorious for their bloodsucking habits
7. nematodes
9. a special type of bony fish with muscular fins supported by stout bones
11. members of the human family
12. a fluid-filled space separating the digestive tract from the outer body wall
13. lizards, snakes, turtles, and crocodiles
15. roundworms
16. the simplest animals with bilateral symmetry
17. vertebrates with hair
20. clams, mussels, scallops, and oysters
22. segmented worms
23. the study of the relationship between evolution and development
24. type of body symmetry in which only a single plane cuts the animal into mirror images
25. a member of the vertebrate class that includes frogs and salamanders
27. feathered endotherms
28. an embryonic stage in which the embryo is a hollow ball
30. the group that consists of reptiles and mammals
31. animals that absorb heat instead of producing it internally
33. the group of scorpions, spiders, ticks, and mites
35. a gas-filled sac in bony fish
36. the social transmission of knowledge, customs, beliefs, and arts

37. the group of terrestrial vertebrates with four legs
38. the type of nerve cord found in the phylum Chordata
39. the sheet of tissue that secretes the shell of a mollusc
41. the group of placental mammals
43. the simplest of all animals, they lack true tissues
46. arthropods with many legs, one pair per body segment
47. the group of annelids that are marine
48. molluscs with a single, spiraled shell
49. a common annelid that eats its way through soil
50. the strap-like rasping organ of a mollusc
52. animals with a notochord, dorsal hollow nerve cord, pharyngeal slits, and post-anal tail
55. type of body symmetry identical around a central axis
56. the group of pouched mammals, including kangaroos and koalas
57. a body cavity completely lined by tissue derived from mesoderm
58. jellyfish, sea anemones, corals, and hydras
60. animals without backbones
61. arthropods with many legs, two pair per body segment
62. the system in a shark that runs along the sides of the body and is sensitive to pressure
63. group of bony fish with lungs
64. the floating body plan of a cnidarian
66. a flexible, longitudinal rod characteristic of chordate animals
67. an internal organ in a female mammal that is used to nurture the embryo
69. a change of body form
70. a protective flap that covers the gills of fishes
72. gill openings in the pharynx, characteristic of chordate animals
73. the most common type of bony fish
74. a body cavity that is not completely lined by tissue derived from mesoderm

DOWN

2. sea urchins, starfish, and sea cucumbers
4. the group that contains all apes
5. a digestive tube with two openings
6. the most abundant arthropods, they have a three-part body
8. the embryonic stage with a two-layered wall and an opening at one end
10. the primate group that includes monkeys, apes, and humans
14. an external skeleton such as that found in all arthropods
18. animals that produce heat internally
19. the type of fish with a skeleton made of bone
21. animals with jointed legs and exoskeletons
22. eukaryotic, multicellular heterotrophs that obtain nutrients by ingestion
26. the type of egg enclosed in a shell and used by reptiles and birds
29. the type of body that is divided into repeated parts
32. the group of crabs, lobsters, crayfish, shrimp, and barnacles
34. the type of fish with a skeleton made of cartilage
40. the type of mammals that are eutherians
42. the sessile body plan of a cnidarian
44. the group of molluscs that includes squids and octopuses
45. the type of tail found in chordate animals
51. a central digestive compartment with only a single opening
53. the group of chordates that all have a cranium and backbone
54. the type of system in echinoderms that uses water pressure to move tube feet
59. a sexually immature animal
65. a group of egg-laying mammals
68. snails, oysters, squids, octopuses, and clams
71. lorises, pottos, lemurs, tarsiers, and anthropoids

CHAPTER 18

An Introduction to Ecology and the Biosphere

Studying Advice

a. Look out a window at some trees and bushes. What determines where any type of wild plant can and will grow? This chapter begins our consideration of the interaction between living organisms and their environments. It answers some basic questions about why certain plants and animals are restricted to just a few parts of the world and how their numbers are limited. If you enjoy walks through a park or wild areas, this chapter will surely be of interest.

b. This chapter also addresses our impact on our natural world. You live in a time of great environmental concern. Your generation has a chance to make a lasting difference.

Student Media

Activities

Science, Technology, and Society: DDT

Adaptations to Biotic and Abiotic Factors

Aquatic Biomes

Terrestrial Biomes

Water Pollution from Nitrates

The Greenhouse Effect

BLAST Animations

The Greenhouse Effect

Graph It

Forestation Change

Global Fresh Water Resources

Municipal Solid Waste Trends in the U.S.
Atmospheric CO_2 and Temperature Changes
Prospects for Renewable Energy

LabBench

Animal Behavior

MP3 Tutors

Ecological Hierarchy
Global Warming

Process of Science

How Can Sow Bug Responses to Environments Be Tested?
How Do Abiotic Factors Affect the Distribution of Organisms?

Videos

Discovery Channel Video: Rain Forests
Hydrothermal Vent
Tubeworms
Flapping Geese
Prokaryotic Flagella
Ducklings
Chimp Cracking Nut
Chimp Agonistic Behavior
Snake Ritual Wrestling
Wolves Agonistic Behavior
Albatross Courtship Ritual
Blue-footed Boobies Courtship Ritual
Giraffe Courtship Ritual

You Decide

Does Human Activity Cause Global Warming?

Organizing Tables

TABLE 18.1 **Define each of the following levels of ecology described in the textbook. These definitions will be a helpful reference as you read and review the chapter.**

Level of Ecology	Textbook Definition
Organismal Ecology	
Population Ecology	
Community Ecology	
Ecosystem Ecology	

TABLE 18.2 **Describe examples of how organisms are affected by the following abiotic factors.**

Abiotic Factor	Examples of How This Factor Impacts Organisms
Energy source	
Temperature	
Nutrients	
Water in terrestrial environments	
Qualities of water in aquatic systems	

TABLE 18.3 **Compare the properties of a river or stream near its source versus its properties near the lake or ocean where it empties.**

Location in the Stream or River	Water Clarity	Water Temperature	Relative Amount of Phytoplankton	Types of Benthic Animals	Types of Predators
Near its source					
Where it empties into a lake or the ocean					

TABLE 18.4 Compare the aquatic biomes and regions in the table.

Aquatic Biome or Region	Location	Special Characteristics
Photic zone		
Aphotic zone		
Benthic realm		
Wetlands		
Pelagic realm		
Coral reefs		
Twilight zone		
Intertidal zones		
Estuaries		
Hydrothermal vent communities		

TABLE 18.5 Compare the terrestrial biomes in the table.

Terrestrial Biome	Vegetation Type	Typical Climate
Tropical Forest		
Savannas		
Deserts		
Chaparral		
Temperate Grassland		
Temperate Broadleaf Forest		
Coniferous Forest		
Tundra		
Polar Ice		

Content Quiz

Directions: Identify the *one* best answer for the multiple-choice questions. For true/false questions, determine if the statement is true or false. If false, change the underlined word(s) to make the statement true. Finally, add the correct word(s) to the fill-in-the-blank questions to make the statements true.

Biology and Society: Penguins and Polar Bears in Peril

1. Which of the following has *not* happened because of global climate changes?
   A. Winter temperatures have risen in Alaska by 5–6°F.
   B. Permanent Arctic sea ice is shrinking.
   C. Sea ice near the Antarctic Peninsula is diminishing.
   D. Rainfall patterns are shifting.
   E. Volcanic activity worldwide has increased.

2. True or False? Adèlie penguins are experiencing a heavy toll on their eggs and chicks because of increased <u>spring storms</u>.

3. Overwhelming evidence indicates that ____________ enterprises are responsible for the global climate changes that are occurring.

An Overview of Ecology

4. Ecology is the scientific study of interactions between:
   A. organisms and their environment.
   B. different types of animals.
   C. animals and plants.
   D. abiotic factors and the environment.
   E. plants and the organisms that pollinate them.

5. True or False? Many ecologists <u>are able to</u> conduct experiments in the field of ecology.

6. Ecologists gain insight from a(n) ____________ approach of watching nature and recording its structure and processes.

ECOLOGY AND ENVIRONMENTALISM, A HIERARCHY OF INTERACTIONS

Matching: Match the level of ecology on the left to its best description on the right.

_____ 7. organismal ecology

_____ 8. community ecology

_____ 9. ecosystem ecology

_____ 10. population ecology

A. the study of energy flow and the cycling of chemicals among the various biotic and abiotic factors

B. the study of the evolutionary adaptations that enable individual organisms to meet the challenges posed by their abiotic environments

C. the study of the interactions between species that affect community structure and organization

D. the study of factors that affect population density and growth

11. Which one of the following correctly lists the levels of ecology in order of increasingly comprehensive levels?
    A. organismal ecology, community ecology, population ecology, ecosystem ecology
    B. population ecology, organismal ecology, community ecology, ecosystem ecology
    C. ecosystem ecology, community ecology, organismal ecology, population ecology
    D. organismal ecology, community ecology, ecosystem ecology, population ecology
    E. organismal ecology, population ecology, community ecology, ecosystem ecology

12. True or False? The distribution of organisms is limited by the abiotic conditions they can tolerate.

13. The ______________ is the sum of all the planet's ecosystems, or all of life and where it lives.

14. The nonliving chemical and physical factors of the environment are the ______________ component.

15. A basic understanding of the field of ______________ is required to understand environmental issues and plan for better practices.

16. An organism's ______________ is the specific environment where it lives and includes the biotic and abiotic factors of its surroundings.

Living in Earth's Diverse Environments

ABIOTIC FACTORS OF THE BIOSPHERE

17. The patchy nature of the biosphere exists mainly because of:
    A. the types of predatory animals that live in a particular region.
    B. differences in climate and other abiotic factors.
    C. random events due to the chance evolution of a new species in a particular region.
    D. the types of plants that grow in a particular region.

18. Which one of the following is an adaptation to conserve water?
    A. a waxy coating on the leaves of plants
    B. the moist surfaces covering gills and lining the lungs
    C. large and very active skeletal muscles
    D. insects flapping their wings at very rapid rates

19. Which one of the following is *not* an abiotic factor?
    A. wind
    B. storms
    C. temperature
    D. predation
    E. water
    F. sunlight

20. True or False? Most organisms <u>cannot</u> maintain a sufficiently active metabolism at temperatures close to 0°C.

21. True or False? Temperatures above 45°C typically destroy an organism's <u>skeleton</u>.

22. True or False? Most photosynthesis occurs near the <u>bottom</u> of a body of water.

23. True or False? Lack of sunlight is <u>usually</u> the most important factor limiting plant growth for terrestrial ecosystems.

24. The distribution and abundance of photosynthetic organisms depend on the availability of inorganic nutrients such as compounds of ____________ and ____________.

25. At the bottom of the ocean, near hydrothermal vents, are ecosystems powered by ____________ bacteria that derive energy from the oxidation of inorganic chemicals such as hydrogen sulfide.

THE EVOLUTIONARY ADAPTATIONS OF ORGANISMS, ADJUSTING TO ENVIRONMENTAL VARIABILITY

Matching: Match the term on the left to its example on the right.

_____ 26. immediate physiological response

_____ 27. anatomical acclimation

_____ 28. physiological acclimation over many days

_____ 29. behavioral response

A. the gradual growth of a heavy coat of fur on a rabbit as winter approaches

B. the production of extra blood cells after a person moves to a higher elevation

C. the sudden appearance of goose bumps as a cold winter wind blasts your skin

D. a bird migrating great distances to reach the best place to feed

30. True or False? Evolutionary adaptation via natural selection results from the interactions of organisms with their <u>environments</u>.

31. True or False? In general, <u>animals</u> are more anatomically changeable than <u>plants</u>.

32. Physiological responses that are reversible but take days or weeks to complete are examples of _____________.

33. The flagging of the tree in Figure 18.12 is an example of a(n) _____________ response.

Biomes

FRESHWATER BIOMES

Matching: Match the term on the left to its best description on the right.

_____ 34. photic zone

_____ 35. aphotic zone

_____ 36. benthic zone

_____ 37. benthos

A. the substrate at the bottom of all aquatic biomes

B. communities of organisms that live in the benthic zone

C. the area where light levels are too low for photosynthesis

D. shallow water near shore and the upper stratum of water away from shore

38. Which one of the following types of biomes occupies the largest part of the biosphere?
   A. savannas
   B. estuaries
   C. grasslands
   D. temperate deciduous forest
   E. aquatic biomes

39. Which one of the following traits is *not* characteristic of a river where it runs into a lake or the ocean?
   A. slow-moving water
   B. warmer water (compared to the water at the source of the river)
   C. clearer water (compared to the water at the source of the river)
   D. predators such as catfish that find food more by scent and taste than by sight
   E. benthic worms and insects that burrow into the muddy bottom

40. Which one of the following statements about wetlands is *false*? Wetlands are:
   A. covered with water either permanently or periodically.
   B. among the lowest of biomes in terms of species diversity.
   C. support the growth of aquatic plants.
   D. provide water storage areas that reduce flooding.
   E. improve water quality by trapping pollutants.

41. True or False? Lakes and ponds that receive large inputs of nitrogen and phosphorus often have <u>low</u> populations of algae.

42. True or False? A stream or river near its source is usually <u>colder</u> and <u>faster</u> than where it reaches a lake or the ocean.

MARINE BIOMES

43. Which one of the following is *not* characteristic of intertidal zones?
   A. daily alternations of submergence in seawater and exposure to air
   B. relatively even temperatures
   C. strong wave action
   D. organisms that can burrow or that can cling to rocks or vegetation

44. Along continental shelves, the photic zone includes:
   A. the pelagic realm.
   B. pelagic and benthic regions.
   C. the twilight zone.
   D. low concentrations of phytoplankton.

45. True or False? Estuaries are crucial feeding areas for <u>many birds</u>.

46. True or False? Many fish and invertebrates use estuaries <u>as breeding grounds</u>.

47. True or False? The ocean's main photosynthetic producers are <u>zooplankton</u>.

48. True or False? The coral reef biome occurs in the <u>pelagic</u> zone of warm tropical waters.

49. In the pelagic photo zone are many free-floating animals called ______________.

50. The environment where a freshwater stream or river merges with the ocean is called a(n) ______________.

51. Large fish, great schools of squid, and whales spend most of their time in the ______________ zone.

52. The most extensive part of the biosphere is the dark ______________ zone, occupying the deepest parts of the ocean.

53. Unusual ______________ communities are powered by chemical energy from Earth's interior instead of sunlight.

HOW CLIMATE AFFECTS TERRESTRIAL BIOME DISTRIBUTION

54. The two main factors that determine the type of biome in a particular region are:
    A. temperature and rainfall.
    B. wind patterns and light intensity.
    C. periodic disturbances and soil quality.
    D. predation and proximity to large bodies of water.

55. Earth's global climate patterns are largely the result of:
    A. magnetic forces and the direction of particular ocean currents.
    B. the gravitational effects of the moon and radiation from outer space.
    C. the movement of geological plates on the surface of the Earth.
    D. radiant energy from the sun and the planet's movement in space.

56. True or False? If the climate in two geographically separate areas is similar, <u>the same</u> type of biome may occur in both.

57. Located primarily around the equator, ______________ are concentrated in the tropics.

58. Latitudes between the tropics and the Arctic Circle in the north and the Antarctic Circle in the south are called ______________ zones.

59. Oceans and large lakes moderate the climate by absorbing ______________.

TERRESTRIAL BIOMES

Matching: Match the term on the left to its best description on the right.

_____ 60. chaparral

_____ 61. savanna

_____ 62. polar ice

_____ 63. tropical rain forest

_____ 64. temperate broadleaf forest

_____ 65. temperate grassland

_____ 66. desert

_____ 67. tundra

_____ 68. coniferous forest

_____ 69. temperate rain forest

A. includes the taiga, the largest terrestrial biome on Earth

B. climate results from location near a cool ocean, also known as the "Mediterranean" biome

C. extreme cold year-round, with low precipitation

D. mostly grasses and scattered trees, warm year-round

E. driest of all biomes, great daily temperature fluctuation

F. dominated by grasses, mostly treeless, periodic fires, and grazing by large mammals

G. equatorial regions with warm and long days year-round

H. occurs at midlatitudes, dense stands of deciduous trees

I. coniferous forests with warm, moist air from the Pacific

J. large areas of the Arctic between the taiga and polar ice

Matching: Match the term on the left to a key feature on the right.

_____ 70. deserts

_____ 71. savannas

_____ 72. tundra

_____ 73. polar ice

_____ 74. tropical forests

A. permafrost is common

B. found at the poles of the Earth

C. large forest canopy, rainfall up to 400 cm a year

D. home to lions, zebras, and antelope in Africa

E. cacti are abundant, water conservation is important

75. Terrestrial ecosystems are grouped into biomes primarily on the basis of their:
    A. type of climate.
    B. type of vegetation.
    C. temperature.
    D. levels of humidity.
    E. physical position on Earth.

76. True or False? The same type of biome throughout the world does not necessarily have the same species living in it.

77. Scientists use a ____________ to visually represent the differences in precipitation and temperature ranges that characterize terrestrial biomes.

78. Species living in the same biome in different parts of the world may look similar to each other because of ____________ evolution.

THE WATER CYCLE

79. Which of the following move from the land to the sea by the global water cycle?
    A. silt
    B. pesticides
    C. chemicals from industrial wastes
    D. All of the above.
    E. None of the above.

80. True or False? Over the oceans, evaporation exceeds precipitation.

81. All parts of the biosphere are linked by the global ____________.

HUMAN IMPACT ON BIOMES

82. Forests of the world are being lost to:
    A. logging.
    B. mining.
    C. agriculture.
    D. air pollution.
    E. All of the above.

83. True or False? Most of the land that we have appropriated is used for housing.

84. The goal of developing, managing, and conserving Earth's resources in ways that meet the needs of people today without compromising the ability of future generations to meet their needs defines ____________.

85. More than the damage to terrestrial ecosystems, the impact of human activities on ____________ ecosystems may pose an even greater threat to life on Earth.

Global Climate Change

THE GREENHOUSE EFFECT AND GLOBAL WARMING

86. Which of the following is most like the affect of greenhouse gases on global temperature?
    A. a person cooling off by swimming in a lake
    B. the inside of a car heating up when parked in the sun
    C. baking a cake in an oven
    D. frying bacon in a pan
    E. microwaving popcorn

87. True or False? Global warming is <u>greater</u> over land than sea.

88. The warming of the atmosphere caused by CO_2, methane, and other gases that absorb heat radiation and slow its escape from Earth's surface is called the ____________.

THE ACCUMULATION OF GREENHOUSE GASES

89. Levels of CO_2 will increase in the global atmosphere if:
    A. global levels of photosynthesis increase.
    B. levels of cellular respiration on Earth are decreased.
    C. we increase the number of plants on Earth.
    D. we continue to rely upon fossil fuels.
    E. None of the above.

90. True or False? The oceans act like massive sponges, soaking up more <u>oxygen</u> than they release.

91. As global temperatures rise, the range of many species will shift <u>toward</u> the poles or to higher elevations.

92. As carbon dioxide is absorbed by the ocean, it makes the oceans more ____________, which will make it more difficult for corals, many species of plankton, and mollusks to build their exoskeletons or shells.

93. As global temperatures increase, organisms living at the tops of mountains may ____________.

EFFECTS OF CLIMATE CHANGE ON ECOSYSTEMS

94. Forest ecosystems in western North America have been hurt by:
    A. bark beetles that more easily attack drought-stressed trees.
    B. bark beetles that can now reproduce twice a year.
    C. decreased levels of water from mountain streams.
    D. longer and more widespread wildfires.
    E. All of the above.

95. True or False? Melting permafrost is shifting the boundary of the tundra <u>southward</u>.

96. Species with reproductive seasons triggered by temperature are moving out of sync with species that use ____________ as an environmental cue for activity.

LOOKING TO OUR FUTURE

97. Which one of the following will *increase* your carbon footprint?
    A. recycling more
    B. eating more beef
    C. riding a bike or walking instead of driving
    D. using less electricity
    E. All of the above will increase your carbon footprint.

98. True or False? Bacteria that live inside cattle or thrive on their manure account for about 20% of <u>methane</u> emissions in the United States.

99. The largest human-related sources of methane are ____________.

100. The amount of greenhouse gas emitted as the result of the actions of a single individual is that person's ____________ footprint.

Evolution Connection: Climate Change as an Agent of Natural Selection

101. Organisms with which of the following traits are most likely to adapt to global climate change?
    A. low genetic variability and short life spans
    B. low genetic variability and long life spans
    C. high genetic variability and short life spans
    D. high genetic variability and long life spans

102. True or False? The rate of climate change is <u>incredibly fast</u> compared to major climate shifts in evolutionary history.

103. If climate change continues, as many as 30% of all plant and animal species will face ____________.

Word Roots

a = without; **bio** = life (abiotic factors: the nonliving chemical and physical factors in an environment)

benth = the depths of the seas (benthic realm: a seafloor or the bottom of a freshwater pond, lake, river, or stream)

bio = life; **sphere** = a ball (biosphere: the sum of all of Earth's ecosystems)

eco = house (ecology: the scientific study of the interactions between organisms and their environments)

inter = between (intertidal zone: the type of zone where land meets the sea)

phyto = a plant; **plankto** = drift or wander (phytoplankton: algae and photosynthetic bacteria that drift passively in aquatic environments)

zo = animal (zooplankton: animals that drift in aquatic environments)

Key Terms

abiotic factors
acclimation
aphotic zone
benthic realm
biome
biosphere
biotic factors
carbon footprint
chaparral
community
community ecology
coniferous forests
coral reef
deserts
ecology
ecosystem
ecosystem ecology
estuary
greenhouse gases
greenhouse effect
habitat
intertidal zone
organismal ecology
pelagic realm
permafrost
photic zone
phytoplankton
polar ice
population
population ecology
savannas
sustainability
taiga
temperate broadleaf forests
temperate grasslands
temperate rain forests
temperate zones
tropical forests
tropics
tundra
wetland
zooplankton

Crossword Puzzle

Use the Key Terms list from this chapter to fill in the crossword puzzle.

ACROSS

3. a terrestrial biome characterized by warm temperatures year-round
4. a terrestrial biome characterized by bitterly cold temperatures
6. the type of ecology that studies how members of a population interact with their environment
9. an ecosystem intermediate between an aquatic one and a terrestrial one
10. a terrestrial biome that includes regions of extremely cold temperature and low precipitation located at high latitudes
14. the type of zone between the tropics and Arctic Circle or the tropics and the Antarctic Circle
15. environment in which an organism lives
16. the diverse algae and cyanobacteria that drift passively in the pelagic zone
17. a terrestrial biome dominated by grasses and scattered trees
18. the northern coniferous forest, characterized by long, snowy winters and short, wet summers
20. the type of realm that includes the open ocean
21. the type of ecology concerned with evolutionary adaptations of individual organisms
24. the type of factor in the environment that is alive
25. all the organisms in a given area, along with the nonliving factors with which they interact
28. a terrestrial biome limited to coastal regions adapted to fire
30. a type of factor in the environment that is not alive
31. all the organisms living together and potentially interacting in a particular area
32. a major type of ecosystem that covers a large geographic region
33. the type of ecology concerned with interactions between species
35. the type of zone that includes water that is not exposed to light
36. an area where fresh water merges with seawater
37. the global ecosystem
38. the type of ecology concerned with energy flow and the cycling of chemicals among the various biotic and abiotic factors

DOWN

1. the study of interactions between organisms and their environments
2. a group of individuals of the same species living in a particular geographic area
5. the goal of developing, managing, and conserving Earth's resources in ways that meet the needs of people today without compromising the ability of future generations to meet their needs
7. animals that drift in the pelagic zone of an aquatic environment
8. continuously frozen subsoil found in the arctic tundra
11. the warming of the atmosphere caused by CO_2, CH_4, and other gases that absorb heat radiation
12. any of the gases in the atmosphere that absorb heat radiation
13. the type of zone where land meets the sea
19. the type of zone, which includes water near shore and at the surface, exposed to light
22. a terrestrial biome characterized by conifers
23. a terrestrial biome characterized by low and unpredictable rainfall
26. the amount of greenhouse gas emitted because of the actions of a person, nation, or other entity
27. a warm-water tropical biome
29. reversible, long-term physiological responses to the environment
32. the type of realm that includes the substrate of a lake, pond, or seafloor
34. the region from the Tropic of Cancer to the Tropic of Capricorn

1
2
3
4
5
6
7
8
9
10
11
12
13
14
15
16
17
18
19
20
21
22
23
24
25
26
27
28
29
30
31
32
33
34
35
36
37
38

CHAPTER 19

Population Ecology

Studying Advice

Have you ever noticed that in some years, there seems to be a lot more flies, mosquitoes, squirrels, birds, deer, or fish? What limits the number of individuals in a particular region? How do we determine how many animals can be killed by hunting and fishing, without wiping out the wild populations? How quickly can a population recover after a disaster such as a horrible storm or fire? This chapter helps to answer these fundamental questions you might have wondered about as you have thought about our natural world.

Student Media

Activities

Techniques for Estimating Population Density and Size
Investigating Survivorship Curves
Madagascar and the Biodiversity Crisis
Introduced Species: Fire Ants
Science, Technology, and Society: DDT
Human Population Growth
Analyzing Age-Structure Pyramids

Biology Labs On-Line

PopulationEcologyLab
DemographyLab

BLAST Animations

Population Dynamics

Graph It

Global Fisheries and Overfishing
Age Pyramids and Population Growth
Municipal Solid Waste Trends in the U.S.

Videos

Discovery Channel Video: Introduced Species
Discovery Channel Video: Emerging Disease

You Decide

Can We Prevent Species Extinction?

Organizing Tables

TABLE 19.1 Compare the Type I, Type II, and Type III survivorship curves in the table.

	Type I	Type II	Type III
Death rate of young			
Death rate of mature adults			
Opportunistic or equilibrial life history?		Intermediate	
Examples			

TABLE 19.2 Compare the opportunistic and equilibrial life histories in the table.

	Opportunistic	Equilibrial
Age of Maturity		
Life Span		
Death Rate		
Number of Offspring		
Parental Care		
Stability of the Population		
Examples		

Content Quiz

Directions: Identify the *one* best answer for the multiple-choice questions. For true/false questions, determine if the statement is true or false. If false, change the underlined word(s) to make the statement true. Finally, add the correct word(s) to the fill-in-the-blank questions to make the statements true.

Biology and Society: Multiplying Like Rabbits

1. Which one of the following statements (A–D) about the introduction of European rabbits and European red foxes into Australia, if any, is *false*? The introduction of European rabbits and European red foxes into Australia:
   A. destroyed farm land.
   B. destroyed grazing land.
   C. led to soil erosion.
   D. let to the extinction of native birds and small mammals.
   E. All of the above statements are true.

2. True or False? Eradication of the European rabbits and European red foxes is impossible.

3. In Australia, European rabbits and European red foxes are _____________ species.

An Overview of Population Ecology

4. Which of the following is *not* a main focus of population ecology?
   A. population size
   B. population density
   C. population structure
   D. predation
   E. growth rate

5. True or False? A community is a group of individuals of a single species occupying the same general area.

6. The study of factors that affect population growth is called _____________.

7. Studying the population ecology of pests and pathogens is an example of _____________ research.

POPULATION DENSITY, POPULATION AGE STRUCTURE

Matching: Match the term on the left to its best description on the right.

_____ 8. population ecology

_____ 9. population age structure

_____ 10. population density

_____ 11. population

A. a group of individuals of the same species living in a given area at a given time

B. the field that examines the factors that influence a population's size, density, and characteristics

C. the distribution of individuals among age groups

D. the number of individuals of a species per unit area or volume

12. Counting the number of bird nests in a particular region would be a way to estimate the population:
    A. age.
    B. density.
    C. structure.
    D. growth rate.

13. The age structure of a population reveals the:
    A. major predators of a population.
    B. main food source of a population.
    C. distribution of individuals among age groups.
    D. types of diseases and reasons for death in a population.

14. True or False? When estimating populations, the <u>larger</u> the number and size of sample plots, the more accurate the estimates of population size.

15. In most cases, it is impractical or impossible to count all the _____________ in a population.

LIFE TABLES AND SURVIVORSHIP CURVES

16. The survivorship curve for humans has:
    A. the highest mortality later in life.
    B. the highest mortality in the middle of the potential life span.
    C. the highest mortality earliest in life.
    D. an even mortality rate throughout the life span.
    E. high rates of mortality early and late in life and low mortality in the middle of the life span.

17. True or False? Oysters have a Type III survivorship curve with the <u>highest</u> mortality early in life.

18. Survivorship and mortality in a population can be tracked by a(n) _____________.

19. A plot of the number of people still alive at each age in a life table is called a(n) ____________.

20. A species that has about an even chance of death at any point in its life would have a Type ____________ survivorship curve.

LIFE HISTORY TRAITS AS EVOLUTIONARY ADAPTATIONS

21. Populations that exhibit an equilibrial life history:
   A. have a Type III survivorship curve.
   B. are mostly smaller-bodied species such as insects.
   C. mature later.
   D. produce many offspring.
   E. do not care for their young.

22. In an unpredictable environment, in which the adult may only have one chance at reproduction, we would expect that these adults would:
   A. have a high number of offspring.
   B. reproduce at a late age.
   C. choose not to reproduce at all.
   D. provide extensive parental care.

23. True or False? Life history traits, like anatomical features, <u>are shaped</u> by adaptive evolution.

24. Organisms that exhibit a(n) ____________ life history tend to grow exponentially when conditions are favorable.

25. Traits that affect an organism's schedule of reproduction and death make up its ____________.

26. True or False? We expect that a female frog that produces thousands of eggs in the spring has <u>extensive</u> parental care.

Population Growth Models

THE EXPONENTIAL POPULATION GROWTH MODEL,
THE LOGISTIC POPULATION GROWTH MODEL

27. Exponential growth typically produces a curve shaped like the letter ____________, whereas logistic growth typically produces a curve shaped most like the letter ____________.
   A. S; J
   B. J; S
   C. L; S
   D. U; J
   E. S; L

28. Examine Figure 19.6. From 1945 to 1950, the male fur seal population:
    A. declined.
    B. reached the carrying capacity.
    C. increased.
    D. experienced exponential growth.
    E. crashed.

29. True or False? In the logistic model, growth rate <u>increases</u> as the population size approaches the carrying capacity.

30. True or False? The exponential model predicts that the larger a population becomes, the <u>faster</u> it grows.

31. The logistic model predicts that a population's growth rate will be highest when the population is at an <u>intermediate</u> level relative to the carrying capacity.

32. Environmental factors that restrict population growth are called ______________ factors.

33. Upon introduction into Australia, the growth of the rabbit population was most like the ______________ growth model.

34. The carrying capacity for a population varies, depending on the species and the ______________ available in the habitat.

REGULATION OF POPULATION GROWTH

35. Which one of the following is most likely a density-dependent factor?
    A. hurricane damage to the nesting sites of birds
    B. a long, very cold winter that kills many bison
    C. a shortage of good browsing vegetation because of overeating
    D. a limited number of nesting sites for birds because of forest fires

36. Aphids and many other insects often show exponential growth in the spring and then rapid die-offs. These populations are:
    A. crashing when they reach their carrying capacities.
    B. limited by factors closely related to population density.
    C. crashing when they are overcome by disease.
    D. most likely limited by density-independent factors.

37. A population that grows exponentially, but is eventually limited by density-dependent factors, will have a growth chart most similar to the shape of the letter:
    A. J.
    B. U.
    C. L.
    D. S.
    E. V.

38. The boom-and-bust cycles of snowshoe hares are most likely caused by:
   A. competition with other grazing mammals.
   B. fluctuations in the lynx population.
   C. the effects of predation.
   D. fluctuations in the hare's food sources.
   E. the combined effects of predation and fluctuations in the hare's food sources.

39. True or False? A density-dependent factor intensifies as the population <u>decreases</u> in size.

40. True or False? Density-independent factors are <u>unrelated</u> to population density.

41. True or False? In many populations, <u>density-independent factors</u> limit population size before density-dependent factors have their greatest impact.

42. A population that remains near its carrying capacity for a long period is most likely limited by ______________ factors.

43. The reliance of individuals of the same species on the same limited resources is called ______________.

Applications of Population Ecology

CONSERVATION OF ENDANGERED SPECIES, SUSTAINABLE RESOURCE MANAGEMENT

44. To achieve the greatest productivity from a population, a resource manager should harvest the population:
   A. down to the lowest levels.
   B. down to about half the carrying capacity.
   C. to about 10% below the carrying capacity, leaving 90% to reproduce.
   D. every third year.

45. Compared to fish on the continental shelves, fish in deeper waters:
   A. grow more slowly and mature more slowly.
   B. grow more slowly but mature more quickly.
   C. grow faster and mature faster.
   D. grow faster but mature more slowly.

46. True or False? <u>Controlled fires</u> were needed to help maintain mature pine trees and to help the red-cockaded woodpecker recover from near-extinction.

47. True or False? Fish populations in deeper waters would be expected to recover <u>faster</u> than fish populations along the shallow continental shelves.

48. A(n) _____________ species is in danger of extinction throughout all or a significant portion of its range.

49. A(n) _____________ species is likely to become endangered in the foreseeable future throughout all or a significant portion of its geographic range.

INVASIVE SPECIES

50. Which one of the following is *not* a characteristic of cheatgrass? Cheatgrass:
    A. fires are more intense and occur more frequently than fires of native plants.
    B. seeds sprout during fall rains and roots continue to grow through the winter.
    C. produces seeds later than native plants.
    D. produces more seeds than its native competitors.
    E. deprives native species or crops of soil moisture and mineral nutrients.

51. True or False? Invasive species typically exhibit an <u>equilibrial</u> life history pattern.

52. A non-native species that has spread far beyond the original point of introduction and causes environmental or economic damage by colonizing and dominating suitable habitats is a(n) _____________ species.

BIOLOGICAL CONTROL OF PESTS

53. Which one of the following is an example of coevolution?
    A. Populations of native grasses are burned regularly by wildfires.
    B. The seeds of plants are spread by flooding along a river.
    C. Hummingbird beak shapes match the shape of the flowers they visit for food.
    D. The shape of a hawk's wing permits soaring, gliding, and swooping to capture prey.

54. Kudzu:
    A. is a native species of North America.
    B. has no natural enemies in the United States.
    C. has been effectively controlled in most parts of the United States using a moth that eats kudzu.
    D. can be killed by a fungal pathogen currently being tested for widespread use.
    E. None of the above is correct.

55. True or False? The invasive Australian rabbit population has been controlled by the introduction of <u>hawks</u> that kill the rabbits.

56. The intentional release of a natural enemy to attack a pest population is called _____________.

INTEGRATED PEST MANAGEMENT

57. Which of the following, if any, is *not* a consequence of the use of pesticides on crops?
    A. Pesticide resistance develops.
    B. Pesticides kill natural predators.
    C. Pesticides kill pollinators.
    D. Pesticides can be carried over great distances as dangerous pollutants.
    E. All of the above are negative consequences of the use of pesticides on crops.

58. True or False? Most crop pests have an <u>opportunistic</u> life history pattern.

59. Typical crops are grown in groups of genetically similar individuals called a(n) ______________.

60. Integrated pest management advocates tolerating low levels of ______________.

Human Population Growth

THE HISTORY OF GLOBAL POPULATION GROWTH

61. Human population growth:
    A. is closest to the model for exponential growth.
    B. is closest to the model for logistic growth.
    C. has remained steady over the course of human history.
    D. is now on the decline, due to density-dependent factors.
    E. has experienced a pattern of boom and bust.

62. A unique feature of human population growth is:
    A. the independence of factors affecting the birth rates.
    B. the independence of factors affecting the death rates.
    C. our ability to voluntarily control it.
    D. that it cannot be limited by density-dependent factors.
    E. that it cannot be limited by density-independent factors.

63. True or False? Since the Industrial Revolution, exponential growth of the human population has resulted mainly from <u>an increase</u> in death rates.

64. Human population growth is based on the same two general parameters that affect other animal and plant populations: ______________ rates and ______________ rates.

AGE STRUCTURES

65. A population that is growing rapidly will have an age structure most like a(n):
    A. square.
    B. inverted pyramid.
    C. pyramid.
    D. circle, widest in the middle and narrowest at the top and bottom.

66. If trends continue as expected, which one of the following age groups in the United States will increase the most in the next 30 years? People who are:
    A. over 80 years old.
    B. 50–55 years old.
    C. 30–35 years old.
    D. 10–20 years old.
    E. 0–4 years old.

67. True or False? The population of the United States <u>continues to increase</u>.

68. The proportion of individuals in different age groups defines the ____________ of that population.

69. The increased proportion of women of childbearing age in the population is known as ____________.

OUR ECOLOGICAL FOOTPRINT

70. In the United States, each person has an ecological footprint about the size of:
    A. a big blanket.
    B. a tennis court.
    C. a basketball court.
    D. a football field.
    E. 18 football fields.

71. True or False? The ecological impact of affluent nations such as the United States is <u>overpopulation</u>.

72. True or False? The world's richest countries, with 20% of the global population, use <u>86%</u> of the world's resources.

73. An estimate of the amount of land required to provide the raw materials an individual or a population consumes defines a(n) ____________.

Evolution Connection: Humans as an Invasive Species

74. Which of the following species were alive in North America 15,000 years ago?
    A. pronghorn antelope
    B. saber-toothed cats
    C. lions
    D. American cheetahs
    E. All of the above.

75. True or False? The speed of changes in the biotic and abiotic environment in North America 10,000 years ago caused the widespread extinction of many large mammal species.

76. The spread of humans over North America about 10,000 years ago had the same impact as many ____________ species today, such as kudzu and starlings.

Word Roots

co = together (coevolution: the evolution of two species in response to each other)
eco = house (ecosystem: an ecological system)
equ = same (equilibrial life history: the type of life history pattern of reaching sexual maturity slowly and producing few offspring but caring for the young)
intra = within (intraspecific competition: competition within a species)

Key Terms

age structure
biological control
carrying capacity
coevolution
density-dependent factor
density-independent factor
ecological footprint
endangered species
equilibrial life history
exponential population growth
intraspecific competition
invasive species
life history
life table
limiting factors
logistic population growth
opportunistic life history
population
population density
population ecology
population momentum
survivorship curve
threatened species

Crossword Puzzle

Use the Key Terms list from this chapter to fill in the crossword puzzle.

ACROSS

2. a plot of the number of people still alive at each age
6. the continuation of population growth in a population reproducing at the replacement rate, as girls in the pre-reproductive age group reach their reproductive years
10. the traits that affect an organism's schedule of reproduction and death
11. a listing of survival and death in a population in a particular time
12. the rate of expansion of a population under ideal conditions
14. a type of species that is in danger of extinction throughout all or a significant portion of its range
18. the number of individuals in a population that an environment can sustain
20. a characteristic of a population referring to the proportion of individuals in different age groups
21. the type of life history pattern of reaching sexual maturity slowly and producing few offspring but caring for the young
22. a type of population-limiting factor that intensifies as the population increases in size
23. a type of population-limiting factor whose intensity is unrelated to population size

DOWN

1. the number of individuals of a species per unit area or volume
3. a type of species that is likely to become endangered in the near future
4. a group of individuals of the same species living in a particular geographic area
5. the type of ecology that studies how members of a population interact with their environment
7. a non-native species that has spread far beyond the original point of introduction
8. the type of competition between individuals of the same species
9. the intentional release of a natural enemy to attack a pest population
13. the type of environmental factor that restricts the number of individuals that can occupy a particular habitat
15. an estimate of the amount of land required to provide the raw materials an individual or a population consumes
16. the type of life history in which organisms reproduce when young and produce many offspring that receive little or no parental care
17. the type of mathematical growth model of idealized population growth that is restricted by limiting factors
19. the reciprocal evolutionary influence between two species

1
2
3
4
5
6
7
8
9
10
11
12
13
14
15
16
17
18
19
20
21
22
23

CHAPTER 20

Communities and Ecosystems

Studying Advice

Most of us are interested in conserving our natural environments and willing to make sacrifices to do so. But how shall we concentrate our efforts? Where can we have the greatest impact? This chapter provides a background that helps us find these answers. It may be the most meaningful chapter in the book. As you read, look for the hope and promise of what we can do together.

Student Media

Activities

Introduced Species: Fire Ants
Interspecific Interactions
Food Webs
Primary Succession
Exploring Island Biogeography
Energy Flow and Chemical Cycling
Pyramids of Production
The Carbon Cycle
The Nitrogen Cycle
Water Pollution from Nitrates
Madagascar and the Biodiversity Crisis
Conservation Biology Review

BLAST Animations

Energy Flow
Carbon Cycle
Nitrogen Cycle

Graph It

Global Fisheries and Overfishing
Animal Food Production Efficiency and Food Policy
Forestation Change
Global Fresh Water Resources
Species Area Effect and Island Biogeography

LabBench

Dissolved Oxygen

MP3 Tutors

Energy Flow in Ecosystems

Process of Science

How Are Impacts on Community Diversity Measured?
How Do Temperature and Light Affect Primary Production?
How Are Potential Prairie Restoration Sites Analyzed?

Videos

Discovery Channel Video: Introduced Species
Discovery Channel Video: Leafcutter Ants
Coral Reef
Clownfish and Anemone
Seahorse Camouflage

You Decide

Can We Prevent Species Extinction?

Organizing Tables

TABLE 20.1 **Describe the three main types of biological diversity in the table, noting an example of each.**

	Definition	Example
Genetic Diversity		
Species Diversity		
Ecosystem Diversity		

TABLE 20.2 **Describe the four main causes of the decline of biological diversity in the table, noting an example of each.**

	Definition	Example
Habitat Destruction		
Invasive Species		
Overexploitation		
Pollution		

TABLE 20.3 **Define and give an example of each of the trophic levels noted in the table. Try to use examples that could all be found in one food chain.**

	Definition	Example
Producer		
Primary Consumer		
Secondary Consumer		
Detritivore		
Decomposer		

Content Quiz

Directions: Identify the *one* best answer for the multiple-choice questions. For true/false questions, determine if the statement is true or false. If false, change the underlined word(s) to make the statement true. Finally, add the correct word(s) to the fill-in-the-blank questions to make the statements true.

Biology and Society: Does Biodiversity Matter?

1. Which one of the following does *not* normally occur?
   A. Ecosystems decompose wastes and recycle nutrients.
   B. Wetlands increase the impact of river flooding.
   C. Natural vegetation helps retain fertile soil and prevent landslides.
   D. Ecosystems purify air and water.
   E. Wetlands buffer coastal populations against hurricanes.

2. True or False? The economic value of ecosystem services in the United States is about <u>twice</u> the national product.

3. About 25% of all prescription drugs contain substances derived from ______________ .

The Loss of Biodiversity

GENETIC DIVERSITY, SPECIES DIVERSITY, ECOSYSTEM DIVERSITY

4. Which one of the following statements is *false*?
   A. About 12% of the known bird species are threatened with extinction.
   B. About 20% of the known mammalian species are threatened with extinction.
   C. About 20% of the known freshwater fish species have either become extinct during human history or are seriously threatened.
   D. About 32% of all known amphibian species are either near extinction or endangered.
   E. Over half of all plant species in the United States have gone extinct since humans have been keeping records.

5. Which one of the following is *not* one of the three main components of biodiversity?
   A. genetic variation within each species
   B. the diversity of ecosystems
   C. the variety of species
   D. the number of individuals of a species

6. True or False? Severe reduction in <u>genetic variation</u> threatens the survival of a species.

7. The genetic ______________ within populations of a species is the raw material that makes microevolution and adaptation to the environment possible.

8. Ecosystem ______________ are functions performed by an ecosystem that directly or indirectly benefit people.

CAUSES OF DECLINING BIODIVERSITY

9. Which one of the following is the single greatest threat to biological diversity?
   A. global warming
   B. depletion of the ozone layer
   C. habitat destruction
   D. introduction of non-native species
   E. overexploitation

10. Whales, the American bison, and the Galápagos tortoises are species that have been:
    A. overexploited.
    B. impacted by the introduction of non-native species.
    C. greatly reduced because of the loss of habitat.
    D. poisoned by the introduction of environmental toxins.

11. True or False? Forest and aquatic ecosystems are threatened by <u>acid</u> precipitation.

12. The introduction of non-native or ______________ species often results in the elimination of native species.

Community Ecology

INTERSPECIFIC INTERACTIONS

13. According to the competitive exclusion principle:
    A. as the number of species in a community increases, interspecific competition decreases.
    B. species will share the community resources to decrease interspecific competition.
    C. when interspecific competition increases, population densities decrease.
    D. if the ecological niches of two species are too similar, they cannot coexist in the same place.

14. What will happen if two species in an ecosystem have identical niches?
    A. Both species will be driven to local extinction.
    B. One species will be driven to local extinction.
    C. One species will be driven to local extinction or one of the species may evolve enough to use a different set of resources.
    D. Both species will be driven to local extinction or both species will evolve enough to use a different set of resources.

15. Which one of the following is *not* an example of a parasitic relationship?
    A. mosquitoes and people
    B. aphids and plants
    C. root-fungus associations called mycorrhizae
    D. a leech and a fish
    E. tapeworms and cattle

16. Which one of the following is most like a parasitic relationship?
    A. a person shopping for groceries
    B. a student writing a paper for a course
    C. a sailor using wind to propel a sailboat
    D. a student stealing a book from another student
    E. a physician treating the wound of a patient

17. Which one of the following is most like a mutualistic relationship?
    A. a person collecting and eating nuts that have fallen from the trees
    B. giving someone $20 to help you change a flat tire
    C. a person moving from one apartment to another
    D. a cow eating grass
    E. bird migrations over great distances

18. Which one of the following statements about predation is *false*?
    A. In a predator-prey relationship, the consumer is the predator.
    B. In a predator-prey relationship, the food species is the prey.
    C. In herbivory, the prey is an animal.
    D. Predation is a form of interspecific competition.
    E. Natural selection refines the adaptations of predators and prey.

19. A moth that has a color pattern that makes it appear to blend into its environment is showing:
    A. warning coloration.
    B. cryptic coloration.
    C. herbivory.
    D. mutualism.

20. True or False? Many animals use <u>warning</u> coloration to make it difficult to spot them in their environment.

21. True or False? In the symbiotic relationship termed <u>predation</u>, one organism benefits at the expense of the other.

22. An organism's ______________ includes abiotic factors, other individuals in its population, and populations of other species living in the same area.

23. An assemblage of species living close enough together for potential interaction is called a(n) ____________.

24. Interactions between species are called ____________ interactions.

25. When populations of two or more species in a community rely on similar limiting resources, they may be subject to ____________.

26. An organism's ecological role is the same as its ecological ____________.

27. A toxic animal will often have a ____________ coloration to caution predators.

TROPHIC STRUCTURE

Matching: Match the term on the left to its best description on the right.

_____ 28. detritivore

_____ 29. primary consumer

_____ 30. producer

_____ 31. secondary consumer

A. an organism that uses photosynthesis

B. an animal that eats herbivores

C. scavengers

D. herbivore

32. What property of a community determines the passage of energy and nutrients from plants and other photosynthetic organisms to herbivores and then to carnivores?
   A. stability
   B. trophic structure
   C. diversity
   D. prevalent form of vegetation

33. Which of the following *never* function as producers in an ecosystem?
   A. terrestrial plants
   B. fungi
   C. phytoplankton
   D. multicellular algae and aquatic plants

34. All organisms in trophic levels above producers are:
   A. autotrophic producers.
   B. autotrophic consumers.
   C. heterotrophic producers.
   D. heterotrophic consumers.

35. Examine the food web in Figure 20.18. Which one of the following changes would most decrease the number of hawks?
   A. an increase in the mouse population
   B. an increase in the snake population
   C. an increase in the owl population
   D. an increase in the lizard population

36. Because of biological magnification, which one of the following members of a food chain will be most affected by the introduction of a toxin into an ecosystem?
   A. producer
   B. herbivore
   C. detritivore
   D. primary consumer
   E. secondary consumer

37. True or False? The trophic level that supports all others is the <u>consumer</u>.

38. Ecologists divide the species of an ecosystem into different ____________ based on their main sources of nutrition.

39. The sequence of food transfer from trophic level to trophic level is called a(n) ____________.

40. An ecosystem's main detritivores are ____________ and ____________.

41. The feeding relationships in an ecosystem are usually woven into elaborate food ____________.

SPECIES DIVERSITY IN COMMUNITIES, DISTURBANCES IN COMMUNITIES, ECOLOGICAL SUCCESSION

42. Species diversity of a community has two components:
   A. species richness and relative abundance.
   B. species richness and disturbances.
   C. succession and relative abundance.
   D. disturbances and succession.
   E. species richness and relative abundance.

43. Which one of the following is the most common sequence of appearance of organisms during primary succession?
   A. lichens and mosses, grasses, autotrophic microorganisms, shrubs, and trees
   B. lichens and mosses, grasses, shrubs, and trees, autotrophic microorganisms
   C. autotrophic microorganisms, lichens and mosses, grasses, shrubs, and trees
   D. autotrophic microorganisms, grasses, shrubs, and trees, lichens and mosses

44. Which of the following might happen to a community to cause secondary succession? A disturbance has:
    A. destroyed most of the shrubs and trees but left most microorganisms, lichens, and grasses intact.
    B. destroyed an existing community but left the soil intact.
    C. completely eliminated one level of a community and the soil.
    D. completely destroyed a community and all the components of its environment.
    E. eliminated the abiotic components of a community.

45. True or False? Ecological disturbances are <u>more common</u> than stability in most communities.

46. True or False? Most communities experience a regular change in <u>species diversity</u>.

47. A ____________ species has an impact on its community that is much larger than its total mass or abundance indicates.

48. The process of community change is called ____________.

Ecosystem Ecology

ENERGY FLOW IN ECOSYSTEMS

49. The two key processes of ecosystem dynamics are:
    A. energy flow and chemical recycling.
    B. energy flow and phase changes.
    C. photosynthesis and metabolism.
    D. phase changes and chemical recycling.

50. Which one of the following statements about energy pyramids is true?
    A. Primary consumers form the lowest level of an energy pyramid.
    B. The highest level of an energy pyramid represents the producers.
    C. Most energy pyramids have 10 to 15 levels.
    D. Herbivores usually appear in the second level of an energy pyramid.

51. Examine the killer whale food chain in Figure 20.15. Within this ecosystem, how would the weight of all the killer whales compare to the collective weight of the zooplankton?
    A. The whale weight would be about 100 times greater.
    B. The whale weight would be about the same as the weight of the zooplankton.
    C. The weight of the zooplankton would be about 10 times greater than the whale weight.
    D. The weight of the zooplankton would be about 100 times greater than the whale weight.
    E. The weight of the zooplankton would be at least 1000 times greater than the whale weight.

52. True or False? Energy reaches most ecosystems in the form of _heat_.

53. True or False? Unlike matter, _energy_ cannot be recycled.

54. True or False? Only _about 1%_ of the visible light that reaches producers is converted to chemical energy by photosynthesis.

55. True or False? On average, only _about 10%_ of the energy in the form of organic matter at each trophic level is stored as biomass in the next level of the food chain.

56. True or False? It takes about the same amount of photosynthetic productivity to produce _10 pounds_ of corn as 1 pound of hamburger.

57. The highest level of biological organization is a(n) ______________.

58. The amount of living organic material in an ecosystem is the ______________.

59. The rate at which producers build organic matter is the ecosystem's ______________.

60. In many developing countries, people primarily eat ______________ because they cannot afford more energy-expensive foods.

61. Chemical elements can be recycled between an ecosystem's living community and the ______________ environment.

62. Plants and other producers acquire their carbon, nitrogen, and other chemical elements in inorganic form from the ______________ and ______________.

CHEMICAL CYCLING IN ECOSYSTEMS

63. A chemical's specific route through an ecosystem depends upon the:
    A. particular element and the trophic structure of the ecosystem.
    B. amount of rainfall and the variation of the seasons.
    C. number of producers and consumers.
    D. amount of carbon and nitrogen in the atmosphere.

64. Which one of the following nutrients is *not* very mobile and is mostly cycled locally?
    A. carbon
    B. nitrogen
    C. phosphorus
    D. water

65. True or False? Plants get most of their _carbon_ from ammonium or nitrate.

66. True or False? Some water can bypass the _biotic_ components of an ecosystem and rely completely on geologic processes.

67. True or False? Each chemical that moves through an ecosystem cycles through an <u>abiotic reservoir</u>.

68. Chemical cycles in an ecosystem are also called ____________ cycles because they include biotic and abiotic components.

69. Fertilizers and pesticides are common sources of ____________, which can result in heavy growth of algae and cyanobacteria.

70. The process of nitrogen fixation converts gaseous ____________ to ammonia and nitrates.

71. Geologic processes such as erosion and the weathering of rock also contribute to the ____________ reservoirs.

72. The burning of wood and fossil fuels contributes to the ____________ cycle.

Conservation and Restoration Biology

BIODIVERSITY "HOT SPOTS"

73. The two main goals of conservation biology are to:
    A. preserve individual species and sustain ecosystems.
    B. preserve individual species and create new ecosystems.
    C. identify new species and restore ecosystems.
    D. identify new species and sustain ecosystems.

74. Examine the map of hot spots in Figure 20.37. In general, most hot spots are located:
    A. in the Northern Hemisphere.
    B. in the Southern Hemisphere.
    C. near the poles.
    D. near the equator.

75. True or False? Biodiversity hot spots are also hot spots of <u>extinction</u>.

76. A species found nowhere else is a(n) ____________ species.

CONSERVATION AT THE ECOSYSTEM LEVEL

77. Which one of the following statements about the edges between ecosystems is *false*?
    A. Edges are prominent features of landscapes.
    B. Edges have their own sets of physical conditions.
    C. Edges often have their own type and amount of disturbance.
    D. Edges can have positive and negative effects on biodiversity.
    E. Landscape edges produced by human activities often have more species.

78. Which one of the following statements about movement corridors is *false*? Movement corridors:
   A. promote dispersal of members of a population.
   B. help sustain populations.
   C. are important to species that migrate between different habitats seasonally.
   D. can reduce the spread of disease.

79. True or False? A habitat is a regional assemblage of interacting ecosystems.

80. The application of ecological principles to the study of land-use patterns is called ____________.

81. A narrow strip or series of small clumps of quality habitat connecting otherwise isolated patches is a(n) ____________.

RESTORING ECOSYSTEMS

82. Restoration ecology may involve:
   A. replanting vegetation.
   B. removing dams.
   C. bioremediation.
   D. fencing out non-native animals.
   E. All of the above.

83. True or False? The rerouting of the Kissimmee River into its natural meandering pathway is an example of restoration ecology.

84. The use of living organisms to detoxify polluted ecosystems defines ____________.

THE GOAL OF SUSTAINABLE DEVELOPMENT

85. Sustainable development requires:
   A. continued ecological research.
   B. the application of ecological knowledge.
   C. the connection of the life sciences with the social sciences, economics, and humanities.
   D. All of the above.

86. True or False? The Sustainable Biosphere Initiative is an attempt to acquire the ecological information necessary for the responsible development, management, and conservation of Earth's resources.

87. Sustainable development increases ____________ and improves the human condition.

Evolution Connection: Biophilia and an Environmental Ethic

88. The concept of ____________ reflects a human desire to affiliate with other life in its many forms.

Word Roots

a = without; **bio** = life (abiotic reservoir: the part of an ecosystem where chemicals accumulate or are stockpiled outside of living organisms)
bio = life; **geo** = the Earth (biogeochemical cycles: the various chemical cycles occurring in an ecosystem)
crypt = hidden (cryptic coloration: another name for camouflage)
detrit = wear off (detritivores: decomposers in an ecosystem)
eco = house (ecology: the scientific study of the interactions between organisms and their environments)
herb = grass; **vor** = eat (herbivore: an animal that eats mostly plants and algae)
inter = between (interspecific competition: competition between species)
intra = within (intraspecific competition: the reliance of individuals of the same species on the same limited resources)
mutu = reciprocal (mutualism: a symbiosis that benefits both partners)
omni = all (omnivore: an animal that eats plants and animals)
para = near; **sit** = food (parasite: an organism that benefits at the expense of another)
phili = love (biophilia: the human desire to affiliate with other life in its many forms)
quatr = by fours (quaternary consumer: an organism that eats tertiary consumers)
sym = together; **bio** = life (symbiotic relationship: an interspecific relationship in which one species lives in or on another species)
terti = the third (tertiary consumer: a consumer that eats secondary consumers)
troph = food (trophic structure: the feeding relationships among the various species making up a community)

Key Terms

abiotic reservoir
biodiversity
biodiversity hot spot
biogeochemical cycles
biological magnification
biomass
biophilia
bioremediation
carnivores
chemical cycling
community
competitive exclusion principle
conservation biology
consumers
cryptic coloration
decomposers
detritivores
detritus
disturbances
ecological niche
ecological succession
ecosystem
ecosystem services
endemic species
energy flow
food chain
food webs
herbivores
herbivory
host
interspecific competition
interspecific interactions
keystone species
landscape
landscape ecology
movement corridor
mutualism
nitrogen fixation
omnivores
parasite
predation
primary consumers
primary production
primary succession
producers
pyramid of production
quaternary consumers
relative abundance
restoration ecology
secondary consumers
secondary succession
species diversity
species richness
sustainable development
tertiary consumers
trophic structure
warning coloration

Crossword Puzzle

Use the Key Terms list from this chapter to fill in the crossword puzzle.

ACROSS

3. the number and relative abundance of species in a biological community
4. a type of ecological succession in which a disturbance has destroyed an existing biological community but left the soil intact
6. a type of predator species that reduces the density of the strongest competitors in a community
7. a type of consumer that eats tertiary consumers
9. the type of ecology that examines the application of ecological principles to the study of land-use patterns
10. a type of symbiosis that benefits both species
11. an organism that makes organic food molecules from carbon dioxide and water and other inorganic molecules
12. a regional assemblage of interacting ecosystems
13. an interaction between species in which one species, the predator, eats the other, the prey
14. the consumption of plant parts or algae by an animal
17. functions performed by an ecosystem that directly or indirectly benefit people
20. an animal that eats plants, algae, or autotrophic bacteria
22. the type of competition between populations of two or more species that require similar limited resources
24. an organism that derives its energy from organic wastes and dead organisms; also called decomposer
25. the feeding relationships in an ecosystem
26. organisms that promote the breakdown of organic materials into inorganic ones
27. the rate at which an ecosystem's plants and other producers build biomass, or organic matter
31. a type of consumer that eats secondary consumers
32. an animal that eats other animals
33. the type of coloration, often bright, of animals possessing chemical defenses
35. nonliving organic matter
36. a type of coloration that is a form of camouflage
39. a type of cycling that reuses chemical elements
40. the type of principle that states that populations of two species cannot coexist in a community if their niches are nearly identical
42. a population's role in its community
43. the type of development that produces long-term prosperity of human societies and the ecosystems that support them
47. a human desire to affiliate with other life in its many forms
48. the larger participant in a symbiotic relationship, serving as home and feeding ground to the symbiont
49. an organism that benefits at the expense of the host
51. a network of interconnecting food chains
53. the amount of organic material in an ecosystem
54. all the organisms living together and potentially interacting in a particular area
56. a type of chemical cycling occurring in an ecosystem, involving both biotic and abiotic components
57. the nonbiological location where components of biogeochemical cycles are stored

DOWN

1. a type of consumer that eats primary consumers
2. a field of ecology that develops methods of returning degraded ecosystems to their natural state
5. the types of interactions between species
8. the process that converts gaseous nitrogen to ammonia and nitrates
15. the passage of energy through the components of an ecosystem
16. a force that changes a biological community and usually removes organisms from it
18. a species that has a distribution limited to a specific geographic area
19. all the abiotic factors in addition to the community of species that exists in a certain area
21. the proportional representation of a species in a biological community; one component of species diversity
23. organisms that obtain food by eating plants or by eating animals that have eaten plants
28. a type of ecological succession in which a biological community arises in an area without soil
29. the use of living organisms to detoxify and restore polluted and degraded ecosystems
30. the sequence of food transfer from producers through several levels of consumers in an ecosystem
34. the accumulation of persistent chemicals in the living tissues of consumers in food chains
37. the type of biology that studies ways to counter the loss of biodiversity
38. a series of small clumps or a narrow strip of quality habitat that connects otherwise isolated patches of quality habitat
41. a diagram depicting the cumulative loss of energy with each transfer in a food chain
44. the total number of different species in a community
45. the process of biological community change resulting from disturbance
46. a small geographic area with an exceptional concentration of species
50. an animal that eats both plants and animals
52. Earth's great variety of life
55. a type of consumer that eats plants, algae, or autotrophic bacteria

1
2
3
4
5
6
7
8
9
10
11
12
13
14
15
16
17
18
19
20
21
22
23
24
25
26
27
28
29
30
31
32
33
34
35
36
37
38
39
40
41
42
43
44
45
46
47
48
49
50
51
52
53
54
55
56
57

CHAPTER 21

Unifying Concepts of Animal Structure and Function

Studying Advice

a. This chapter relates to many activities that occur inside your body. Pause occasionally while reading to reflect on events that you have noticed in yourself.

b. Before reading, flip through the chapter and look over the figures. This will give you a feel for what the chapter is about and help you identify figures that you will want to refer to again as you read and review.

c. Complete the following organizing tables as you read, to help refine your studies.

Student Media

Activities

The Levels of Life Card Game
Overview of Animal Tissues
Epithelial Tissue
Connective Tissue
Muscle Tissue
Nervous Tissue
Correlating Structure and Function of Cells
Regulation: Negative and Positive Feedback
Structure of the Human Excretory System
Nephron Function
Control of Water Reabsorption

BLAST Animations

Negative Feedback: Body Temperature
Anatomy of the Kidney
How the Kidney Works

MP3 Tutors

Animal Structure
Kidney Function

Process of Science

How Does Temperature Affect Metabolic Rate in *Daphnia*?
What Affects Urine Production?

Videos

Discovery Channel Video: An Introduction to the Human Body

Organizing Tables

TABLE 21.1 Compare the definitions of the following pairs of terms in the table.

Definition	Definition
Negative feedback:	Positive feedback:
Endotherm:	Ectotherm:
Anatomy:	Physiology:
Open system:	Closed system:
Osmoconformers:	Osmoregulators:
Reabsorption:	Secretion:

TABLE 21.2 Define each of the following types of tissues and provide at least two examples of each.

Tissue Type	Definition/Key Features	Examples
Epithelia		
Connective		
Muscular		
Nervous		

TABLE 21.3 Describe the main events of urine formation.

Step	What happens?
Filtration	
Reabsorption	
Secretion	
Excretion	

Content Quiz

Directions: Identify the *one* best answer for the multiple-choice questions. For true/false questions, determine if the statement is true or false. If false, change the underlined word(s) to make the statement true. Finally, add the correct word(s) to the fill-in-the-blank questions to make the statements true.

Biology and Society: Keeping Cool

1. Which one of the following will *not* help to cool down the body of an exercising athlete?
   A. lowered blood pressure
   B. widened surface blood vessels
   C. decreased activity
   D. evaporation of sweat

2. True or False? Exercising in a hot environment can lead to intense sweating, dehydration, an eventual drop in blood pressure, and then fainting in a condition called heat <u>stroke</u>.

3. Under extreme conditions of intense heat, a person may suffer from ______________, in which the body temperature rises dangerously high, the skin becomes hot and dry, and organs start to fail.

The Structural Organization of Animals

FORM FITS FUNCTION

4. The form/function relationship of biological equipment is a result of:
   A. natural selection.
   B. design.
   C. random events.
   D. purposeful planning.

5. True or False? The principle of "form fits function" applies to all levels of biological organization.

6. The study of the *structure* of an organism and its parts defines ____________, and the study of the *function* of an organism's structural equipment defines ____________.

TISSUES

Matching: Match the tissue on the left to its description on the right.

_____ 7. muscular tissue

_____ 8. epithelial tissue

_____ 9. nervous tissue

_____ 10. connective tissue

A. covers the surface of the body and lines organs and cavities within the body

B. found in the brain and spinal cord, cells in this tissue carry signals very rapidly over long distances

C. a sparse population of cells scattered through an extensive extracellular matrix

D. cells have specialized proteins arranged into a structure that contracts when the cell is stimulated

11. Which one of the following is a connective tissue with a matrix that is liquid instead of solid?
    A. adipose tissue
    B. blood
    C. cartilage
    D. bone
    E. fibrous connective tissue

12. Which one of the following is an involuntary muscle tissue found in the walls of the digestive tract and blood vessels?
    A. smooth muscle
    B. cardiac muscle
    C. skeletal muscle
    D. fibrous muscle

13. True or False? The basic unit of nervous tissue is the spinal cord.

14. The most widespread connective tissue in the vertebrate body is ____________.

ORGANS AND ORGAN SYSTEMS

15. Which one of the following is *not* a body organ?
    A. brain
    B. blood
    C. stomach
    D. liver
    E. lungs

16. True or False? Two or more tissues packaged into one working unit that performs a specific function defines a(n) tissue.

17. Teams of organs that work together to perform a vital bodily function defines ____________.

Exchanges with the External Environment

18. Which one of the following statements is *false*?
    A. In an animal body, every living cell must be bathed in a watery solution.
    B. Exchange with the environment is easier for multicellular animals than single-celled organisms.
    C. Simple body forms, such as two-layered sacs and flat shapes, do not allow much complexity in internal organization.
    D. The epithelium of the human lungs has a very large total surface area.
    E. The digestive, respiratory, and urinary systems exchange materials with the external environment.

19. True or False? Every organism is a(n) closed system because all life continuously exchanges chemicals and energy with its surroundings.

20. The cells of more complex animals indirectly exchange materials with the environment by connecting exchange surfaces with body cells through a(n) ____________ system.

Regulating the Internal Environment

HOMEOSTASIS

21. Which one of the following is the best example of homeostasis?
    A. the elevation of hormones in the blood during puberty
    B. the elevation of hormones in the blood during pregnancy
    C. an increase in body temperature in response to a bacterial infection
    D. the water content of your cells stays about the same no matter what you drink

22. True or False? All of the body's organ systems play a role in homeostasis.

23. True or False? Homeostasis in the human body permits small changes to occur.

24. The body's tendency to maintain relatively constant conditions in the internal environment despite changes in the external environment defines ____________.

NEGATIVE AND POSITIVE FEEDBACK

25. The part of a negative feedback mechanism in a household furnace that monitors temperature and switches the heater on and off is the:
    A. thermostat.
    B. end zone.
    C. set point.
    D. deflector.
    E. heater.

26. Which one of the following is a positive feedback mechanism?
    A. scratching a healing wound that itches, reinjuring the skin, which then itches as it heals
    B. shivering to warm up the body and then stopping after you warm up
    C. eating when you are hungry and stopping when you are full
    D. drinking a beverage to quench a thirst
    E. None of the above is a positive feedback mechanism.

27. True or False? Negative feedback occurs when the results of some process <u>promote</u> that very process.

28. In ____________ feedback, the results of a process encourage that very process.

THERMOREGULATION

29. Which one of the following animals is an endotherm?
    A. butterfly
    B. trout
    C. frog
    D. mouse
    E. turtle

30. A mild fever while fighting an infection:
    A. should be lowered by taking acetaminophen.
    B. discourages bacterial growth.
    C. slows the body's internal defenses.
    D. is a sign of a weak immune system.

31. True or False? Animals that maintain a body temperature substantially warmer than the surrounding environment are called <u>ectotherms</u>.

32. The maintenance of internal body temperature within defined limits defines ____________.

OSMOREGULATION

33. Living cells depend on a precise balance of ____________ for proper osmoregulation.
    A. salts and minerals
    B. chemical and solar energy
    C. heat and cold
    D. sugars and carbohydrates
    E. water and solutes

34. True or False? Most marine invertebrates are <u>osmoconformers</u>.

35. True or False? All land animals are <u>osmoconformers</u>.

36. The control of the gain or loss of water and dissolved solutes in animals defines ____________.

HOMEOSTASIS IN THE URINARY SYSTEM

Matching: Match the property on the left to its best description on the right.

_____ 37. nephron	A. urine leaves the kidneys in this
_____ 38. ureter	B. urine is stored in this
_____ 39. urinary bladder	C. the functional unit of the kidneys
_____ 40. urethra	D. urine leaves the bladder in this

41. The urinary system plays a central role in homeostasis, forming and excreting ____________ while regulating the amount of ____________ in body fluids.
    A. urine; water and salts
    B. blood; sugars
    C. feces; blood
    D. urine; sugars
    E. sweat; water and sugars

42. Which one of the following statements is *false*?
    A. The spleen is the main processing center in the human urinary system.
    B. The urinary system plays a central role in homeostasis.
    C. Every day, the total volume of blood in the body passes into the kidneys hundreds of times.
    D. The body uses hormones to control the internal concentration of water and dissolved molecules.
    E. Humans with one functioning kidney can lead a normal life.

43. Which of the following is the correct sequence of urinary functions performed by the kidneys?
   A. reabsorption, filtration, secretion, excretion
   B. secretion, excretion, reabsorption, filtration
   C. secretion, filtration, excretion, reabsorption
   D. filtration, reabsorption, secretion, excretion
   E. filtration, reabsorption, excretion, secretion

44. True or False? The number of kidneys available for transplant is <u>not enough</u> to meet the current demand for transplants.

45. A person who suffers from kidney failure can be treated by using a blood filtration machine in a process called ____________.

Evolution Connection: Adaptations for Thermoregulation

46. Which one of the following is an adaptation that helps to keep an animal's body warm?
   A. panting
   B. sweating
   C. bathing
   D. shivering

47. True or False? Most land mammals react to cold by <u>compressing</u> their fur.

48. Migration to warmer parts of the world is an example of a(n) ____________ adaptation that promotes thermoregulation.

49. Hair, feathers, or fat layers that increase insulation are examples of ____________ adaptations that promote thermoregulation.

50. Sweating is an example of a(n) ____________ adaptation that helps an animal keep cool.

Word Roots

ecto = outside; **therm** = heat (ectotherms: animals that obtain body heat primarily by absorbing it from their surroundings)

endo = inside (endotherms: organisms whose bodies are warmed by heat generated by metabolism; this heat is used to maintain a body temperature higher than that of the external environment)

epi = upon or over (epithelium: tissues that cover the surfaces of the body)

fibro = a fiber (fibrous connective tissue: a dense tissue with large numbers of collagenous fibers organized into parallel bundles)

homeo = same; **stasis** = standing (homeostasis: the steady-state physiological condition of the body)

inter = between (interstitial fluid: the fluid that fills the space between cells)

nephro = the kidney (nephron: the functional unit of a kidney)

neuro = a nerve (neuron: a nerve cell)

Key Terms

adipose tissue
anatomy
blood
bone
cardiac muscle
cartilage
connective tissue
dialysis
ectotherms
endotherms
epithelial tissue
epithelium
excretion
fever
fibrous connective tissue
filtrate
filtration
homeostasis
interstitial fluid
loose connective tissue
muscle tissue
negative feedback
nephron
nervous tissue
neuron
open system
organ
organ systems
osmoconformers
osmoregulation
osmoregulators
physiology
positive feedback
reabsorption
secretion
skeletal muscle
smooth muscle
thermoregulation
tissue
tubules
ureter
urethra
urinary bladder
urine

Crossword Puzzle

Use the Key Terms list from this chapter to fill in the crossword puzzle.

ACROSS

3. the tubular excretory unit of the vertebrate kidney
5. the most widespread type of connective tissue in the vertebrate body
6. the type of tissue that is capable of contracting
8. the type of feedback in which a physiological variable triggers a response that counteracts the initial fluctuation
9. the process by which water and valuable solutes are reclaimed from the filtrate and returned to the blood
13. the extraction of water and small solutes in the kidney
14. the pouch where urine is stored prior to elimination
15. type of tissue that stores fat
16. type of muscle that forms the contractile wall of the heart
17. a type of tissue that forms sheets that cover and line body surfaces
18. category of animal tissue that functions mainly to bind and support other tissues
20. an abnormally high internal body temperature, usually the result of an infection
21. the separation of small blood molecules in which some are kept and others discarded
22. the type of dense connective tissue that forms tendons
23. an animal that warms itself mainly by absorbing heat from its surroundings
26. the study of the functions of an organism
28. a specialized center of body function composed of several different types of tissues
29. fluid extracted by the excretory system from the blood or body cavity
30. the control of water balance in organisms
31. the type of system in which blood directly bathes the organs
33. a group of organs that work together to perform vital body functions
34. an integrated group of cells with a common structure and function
35. an animal whose body fluids have a different osmolarity than the environment
37. the type of tissue made up of neurons and supportive cells
38. a type of feedback in which a change in some variable triggers mechanisms that amplify the change
39. a thin tube within the internal structure of the human kidney
40. the type of muscle that lacks striations

DOWN

1. the maintenance of internal temperature within a narrow range
2. the type of fluid that fills the spaces between cells
4. another name for a nerve cell
7. a tube that releases urine from the body
10. a type of connective tissue with a fluid matrix called plasma
11. type of connective tissue consisting of living cells held in a rigid matrix of collagen fibers embedded in calcium salts
12. the waste material produced by the vertebrate excretory system
15. the study of the structure of an organism
16. type of flexible connective tissue found at ends of bones
17. the disposal of nitrogen-containing waste products of metabolism
19. the type of muscle responsible for voluntary movements of the body
24. an animal that uses metabolic energy to maintain a constant body temperature
25. the steady-state physiological condition of the body
27. an animal that is isotonic with its environment
32. a duct leading from the kidney to the urinary bladder
36. the discharge of molecules synthesized by a cell

1
2
3
4
5
6
7
8
9
10
11
12
13
14
15
16
17
18
19
20
21
22
23
24
25
26
27
28
29
30
31
32
33
34
35
36
37
38
39
40

CHAPTER 22

Nutrition and Digestion

Studying Advice

a. Do you pay attention to the latest news about how to be healthy? Have you ever struggled to lose weight? Do you know someone who has battled an eating disorder? This chapter is full of information related to these and other interesting questions. Some might even say that this chapter is the reward for learning something about chemistry!

b. Use your natural curiosity as motivation to read this chapter and better understand how to live a healthy life.

Student Media

Activities

How Animals Eat Food
Human Digestive System
Analyzing Food Labels
Case Studies of Nutritional Disorders

MP3 Tutors

The Human Digestive System

Process of Science

What Role Does Amylase Play in Digestion?

Videos

Discovery Channel Video: Nutrition
Shark Eating a Seal
Lobster Mouth Parts
Hydra Eating Daphnia (time-lapse)

You Decide

Is Ephedra Safe and Effective?

Low-fat or Low-carb Diets—Which is Healthier?

Organizing Tables

TABLE 22.1 Compare the definitions and list examples of organisms using each of the following diets.

Type of Diet	Definition	Examples of Organisms Using This Type of Diet
Herbivore		
Carnivore		
Omnivore		

TABLE 22.2 Compare the main events and locations of each of the four stages of food processing.

Stage	Description of This Process	Digestive Compartment(s) Where This Occurs
Ingestion		
Digestion		
Absorption		
Elimination		

TABLE 22.3 Compare the different types of essential human nutrients.

Type of Nutrient	Functions in the Body	Number That Are Essential to Humans
Essential amino acids		
Vitamins		
Minerals		

Content Quiz

Directions: Identify the *one* best answer for the multiple-choice questions. For true/false questions, determine if the statement is true or false. If false, change the underlined word(s) to make the statement true. Finally, add the correct word(s) to the fill-in-the-blank questions to make the statements true.

Biology and Society: Stomach Surgeries

1. The most common weight-loss surgery in the United States is:
   A. biliopancreatic diversion.
   B. gastric bypass.
   C. gastric banding.
   D. intestinal diversion.

2. True or False? Most weight-loss surgeries are successful if accompanied by <u>a healthy diet and exercise</u>.

3. About a third of American adults are overweight or _____________.

Overview of Animal Nutrition

Matching: Match the term on the left to its best definition on the right.

_____	4. elimination	A.	eating
_____	5. absorption	B.	the breakdown of food to relatively small nutrient molecules
_____	6. ingestion	C.	the uptake of small nutrient molecules by the body's cells
_____	7. digestion	D.	the disposal of undigested materials from meals

8. Which one of the following ingests plants and animals?
   A. herbivore
   B. omnivore
   C. carnivore
   D. endovore
   E. maximore

9. Food is dismantled inside an animal's body because food molecules:
   A. are too large to cross their cell membranes.
   B. are usually different from molecules that make up an animal's body.
   C. A and B.
   D. None of the above.

10. Gastrovascular cavities are found in:
    A. cnidarians and flatworms.
    B. reptiles and birds.
    C. many single-celled organisms and some multicellular organisms, such as sponges.
    D. herbivores but not carnivores.

11. True or False? <u>Mechanical</u> digestion is the chemical breakdown of food by digestive enzymes.

12. In hydrolysis, polymers are broken down into ____________.

13. The simplest digestive compartments are ____________, intracellular membrane-bound organelles filled with digestive enzymes.

A Tour of the Human Digestive System

Matching: Match each item on the left to its best description on the right.

_____ 14. pepsin

_____ 15. esophagus

_____ 16. *Helicobacter pylori*

_____ 17. gastric juice

_____ 18. small intestine

_____ 19. alimentary canal

_____ 20. duodenum

_____ 21. pancreas

_____ 22. stomach

_____ 23. liver

_____ 24. large intestine

_____ 25. pharynx

_____ 26. microvilli

_____ 27. gallbladder

A. the cause of most stomach ulcers

B. a large sac that can store enough food to sustain us for hours

C. the first section of the small intestine

D. an intersection of the food and breathing pathways

E. where bile is produced

F. a muscular tube that connects the pharynx to the stomach

G. site of water resorption and the formation of feces

H. strong acid, digestive enzymes, and mucus

I. an enzyme that digests proteins

J. another name for the entire digestive tube

K. the longest part of the alimentary canal

L. a gland that secretes pancreatic juice into the duodenum

M. where bile is stored

N. microscopic projections on intestinal cells

28. What keeps the stomach from just eating itself from the inside out?
    A. A protective coating of mucus limits the contact between the cells lining the stomach and the gastric juices.
    B. Nerves and hormones limit the secretion of gastric juice to periods when food is in the stomach.
    C. About once every three days, new cells must replace the stomach lining that is damaged.
    D. All of the above.
    E. None of the above.

29. Today, most ulcers are effectively cured by:
    A. antacids.
    B. calcium supplements.
    C. antibiotics.
    D. milk.
    E. vitamins.

30. True or False? Occasional backflow of acid chyme into the esophagus causes <u>ulcers</u>.

31. The enzyme salivary ______________ hydrolyzes starch.

32. Food is moved through the esophagus and intestines by ______________, rhythmic waves of muscular contractions that squeeze the food ball along the esophagus.

33. Rings of muscle that control the movement of materials into and out of the stomach are called ______________.

34. The last 6 inches of the large intestine is called the ______________, where feces are stored until they can be eliminated.

Human Nutritional Requirements

35. Cellular respiration uses ______________ to break down sugar and other food molecules, generating many molecules of ______________ for cells to use as a direct source of energy.
    A. oxygen; ATP
    B. carbon dioxide; oxygen
    C. ATP; carbon dioxide
    D. water; oxygen
    E. water; ATP

36. True or False? Dietary calories listed on food labels are actually <u>kilocalories</u>.

37. True or False? Most <u>animal</u> proteins are deficient in one or more of the essential amino acids.

38. True or False? Overdoses of certain <u>vitamins</u> (such as A, D, and K) can be harmful.

39. True or False? Minerals are organic substances required in a healthy diet.

40. The amount of energy it takes just to maintain your body is called the _____________.

41. Essential amino acids must be obtained in the _____________.

42. Most vitamins function as assistants to _____________ in catalyzing metabolic reactions.

43. Calcium and sodium are examples of the 21 different types of inorganic nutrients called _____________ required in the human diet.

44. Cells use cellular respiration to extract _____________ stored in the organic molecules of food and use it to do work.

45. About 60% of our food energy is lost as _____________ that dissipates to the environment.

Nutritional Disorders

46. Which one of the following statements is *false*?
    A. In humans, obesity increases the risk of heart attack, diabetes, and several other diseases.
    B. The best way to maintain a healthy weight is to exercise and eat a balanced diet.
    C. The most reliable sources of essential amino acids are animal products.
    D. Most victims of protein deficiency are adults.
    E. Eating disorders can also cause malnutrition.

47. True or False? Anorexia nervosa is a behavioral pattern of binge eating followed by purging.

48. True or False? We do not know what causes anorexia and bulimia.

49. In Africa, the syndrome named kwashiorkor refers to a disease caused by a deficiency of _____________.

50. About one-third of all Americans are _____________, with an inappropriately high body mass index.

Evolution Connection: Fat and Sugar Cravings

51. Fat cravings likely evolved in our human ancestors because they were always:
    A. in danger of not finding enough water.
    B. in danger of starvation.
    C. running short of protein.
    D. suffering from obesity.
    E. reproducing.

52. True or False? The majority of Americans consume too many fatty foods, which contributes to obesity.

53. For our ancestors, foods that were ____________ or sweet were probably difficult to find.

Word Roots

an = without; **orex** = appetite (anorexia: an eating disorder characterized by self-starvation due to an intense fear of gaining weight)

bulim = hunger (bulimia: a pattern of binge eating followed by purging through induced vomiting)

carni = flesh; **vora** = eat (carnivores: animals that eat other animals)

gastro = stomach; **vascula** = a little vessel (gastrovascular cavities: compartments with a single opening through which food enters and undigested food is expelled)

herb = grass (herbivore: an animal that eats plants)

kilo = a thousand (kilocalorie: a thousand calories)

mal = bad (malnutrition: a deficiency in one or more of the essential nutrients)

omni = all (omnivore: an animal that consumes meat and plant material)

peri = around; **stalsis** = a constriction (peristalsis: rhythmic waves of contraction of smooth muscle that push food along the digestive tract)

Key Terms

absorption
alimentary canal
anorexia nervosa
anus
appendix
basal metabolic rate (BMR)
bile
body mass index (BMI)
bulimia
calorie
carnivores
chemical digestion
chyme
colon
digestion
digestive tubes
duodenum
elimination
esophagus
essential amino acids
essential fatty acids
essential nutrients
feces
food vacuoles
gallbladder
gastric juice
gastrovascular cavities
herbivores
ingestion
kilocalorie
large intestine
liver
malnutrition
mechanical digestion
metabolic rate
minerals
mouth
obesity
omnivores
oral cavity
pancreas
pepsin
peristalsis
pharynx
rectum
salivary glands
small intestine
stomach
tongue
vitamins

Crossword Puzzle

Use the Key Terms list from this chapter to fill in the crossword puzzle.

ACROSS

3. the type of nutrients required for a plant to grow and complete its life cycle
7. the type of amino acids that an animal cannot synthesize
8. an enzyme present in gastric juice that begins the hydrolysis of proteins
9. the passing of undigested material out of the digestive compartment
11. the process of breaking food into molecules small enough for the body to absorb
13. three letter acronym for a ratio of weight to height used as a measure of obesity
14. mixture of recently swallowed food and gastric juice
15. a digestive tract consisting of a tube running between a mouth and an anus
16. an opening into which food is taken into an animal's body
17. opening through which undigested materials are expelled
18. an excessively high body mass index
20. three-letter acronym for the number of kilocalories a resting animal requires to fuel its essential body processes for a given time
21. the wastes of the digestive tract
22. tubular portion of the gut, composed of the colon and rectum
24. eating disorder that results in self-starvation due to an intense fear of gaining weight
28. an animal that eats other animals
29. the type of digestion involving the physical breakdown of food into smaller pieces
30. the organ that produces bile and detoxifies poisonous chemicals in the blood
32. eating disorder characterized by episodic binge eating followed by induced vomiting
33. the type of glands that secrete substances to lubricate food and begin the process of chemical digestion
34. an animal that eats mainly plants or algae
35. energy expended by the body per unit time
36. the absence of one or more essential nutrients from the diet
37. a thousand calories
39. an area in the vertebrate throat where air and food passages cross
41. an organic molecule required in the diet in very small amounts
43. another name for mouth
44. a gland that secretes digestive enzymes, insulin, and glucagon
45. the longest section of the alimentary canal

DOWN

1. an organ that stores bile and releases it as needed into the small intestine
2. rhythmic waves of contraction that push food along the digestive tract
3. the type of fatty acid that animals need but cannot make
4. the simplest type of digestive cavity, found in single-celled organisms
5. a tube that connects the pharynx to the stomach
6. the first section of the small intestine
10. the vertebrate alimentary canal between the small intestine and the rectum
12. a digestive compartment with two openings
13. a mixture of substances stored in the gall bladder
14. the type of digestion involving the chemical breakdown of food
15. the uptake of small nutrient molecules by an organism's own body
19. in nutrition, a chemical element other than hydrogen, oxygen, or nitrogen that an organism requires for proper body functioning
23. the collection of fluids secreted by the epithelium lining the stomach

DOWN

25. a small, fingerlike extension near the union of the small and large intestines
26. the amount of energy that raises the temperature of 1 ml of water by 1 degree Celsius
27. the type of digestive cavity with a single opening for the entrance of food and exit of wastes
31. the act of eating
33. a pouch-like organ in a digestive tube that grinds and churns food and may store it temporarily
38. a heterotrophic animal that consumes meat and plant material
40. the terminal portion of the large intestine where the feces are stored until elimination
42. muscular organ of the mouth that helps to taste and swallow food

1
2
3
4
5
6
7
8
9
10
11
12
13
14
15
16
17
18
19
20
21
22
23
24
25
26
27
28
29
30
31
32
33
34
35
36
37
38
39
40
41
42
43
44
45

CHAPTER 23

Circulation and Respiration

Studying Advice

a. This chapter and the other animal physiology chapters have great relevance to the mechanisms and wonders of your own body. Take the time to notice in yourself the many human structures and processes discussed in these chapters.

b. As you read, create your own simple flowcharts showing the path of blood through the body and the path of air through the lungs. Then keep these flowcharts handy to clarify events discussed elsewhere in the chapter.

Student Media

Activities

Path of Blood Flow in Mammals
Mammalian Cardiovascular System Structure
Mammalian Cardiovascular System Function
The Human Respiratory System
Transport of Respiratory Gases

Biology Labs On-Line

CardioLab
HemoglobinLab

BLAST Animations

Anatomy of the Heart
Cardiac Cycle Overview
Electrical Coordination of the Cardiac Cycle

LabBench

Circulatory Physiology

MP3 Tutors

The Human Circulatory System
Human Respiration

Process of Science

How Is Cardiovascular Fitness Measured?

Videos

Discovery Channel Video: Blood

You Decide

Is Second-hand Smoke Dangerous?

Organizing Tables

TABLE 23.1 Compare open and closed circulatory systems.

	Open Circulatory System	Closed Circulatory System
The path of blood		
Examples of animal groups that have this system		

TABLE 23.2 Compare the functions of the four chambers in a human heart.

Heart Chamber	Receives Blood From	Sends Blood To
Right atrium		
Right ventricle		
Left atrium		
Left ventricle		

TABLE 23.3 Compare the structures and functions of blood vessels.

Type of Blood Vessel	Structure of the Walls	Presence of One-Way Valves?
Capillaries		
Arteries		
Veins		

TABLE 23.4 Compare the structures and functions of the components of human blood.

Component	Structure	Function(s)
Red blood cells		
White blood cells		
Platelets		

TABLE 23.5 Compare the four main types of respiratory surfaces found in animals.

Type of Surface	Description of the Structure	Examples of Animals with This Structure
Skin		
Gills		
Tracheae		
Lungs		

Content Quiz

Directions: Identify the *one* best answer for the multiple-choice questions. For true/false questions, determine if the statement is true or false. If false, change the underlined word(s) to make the statement true. Finally, add the correct word(s) to the fill-in-the-blank questions to make the statements true.

Biology and Society: The ABCs of Saving Lives

1. The three letters in the ABCs of lifesaving stand for:
   A. All Breathing Is Critical
   B. Airway, Breathing, Circulation
   C. Always Be Careful
   D. Attitude, Body, and Control
   E. Airway, Blocked, Is Critical

2. True or False? If a man's <u>heart</u> fails, he may need CPR to survive.

3. The respiratory system works closely with the ____________ system.

Unifying Concepts of Animal Circulation

OPEN AND CLOSED CIRCULATORY SYSTEMS

4. Which one of the following is *not* a main component of a circulatory system?
   A. a vascular system
   B. a circulating fluid
   C. lungs or gills
   D. a central pump

5. Which one of the following is the functional center of the circulatory system?
   A. arteries
   B. capillaries
   C. arterioles
   D. veins
   E. heart

6. True or False? Blood is confined to vessels in <u>an open</u> circulatory system.

7. True or False? Metabolic wastes, such as <u>oxygen</u>, diffuse from cells into the circulatory system.

8. The ____________ system in humans includes the heart and blood vessels.

The Human Cardiovascular System

THE PATH OF BLOOD

9. Which one of the following represents a correct sequence of blood flow through a part of the human cardiovascular system?
   A. left ventricle → pulmonary veins → capillaries in the lungs → pulmonary arteries → right atrium
   B. left ventricle → pulmonary arteries → pulmonary veins → capillaries in the lungs → right atrium

C. right ventricle → pulmonary veins → capillaries in the lungs → pulmonary arteries → right atrium
D. right ventricle → pulmonary arteries → capillaries in the lungs → pulmonary veins → left atrium
E. right atrium → right ventricle → left atrium → capillaries in the lungs → left ventricle

10. True or False? The systemic circuit carries blood from the heart to organs in the rest of the body.

11. Humans and other terrestrial vertebrates have a(n) _____________ circulatory system with two distinct circuits of blood flow.

How the Heart Works

12. Which one of the following statements about the human circulatory system is *false*?
    A. The atria are much more powerful than the ventricles.
    B. In the heart, blood going to the lungs is kept separate from blood going out to the body.
    C. Valves in the heart keep blood moving in the correct direction.
    D. An adult human heart is about the size of an adult human fist.

13. Which one of the following statements about the control of heart rate is *false*?
    A. The SA node generates electrical impulses in the right atrium.
    B. During a heart attack, the pacemaker is often unable to maintain normal rhythm.
    C. The SA node receives an impulse from the AV node.
    D. The pacemaker of the heart is composed of specialized muscle cells.

14. True or False? Blood in the heart moves only in one direction due to one-way valves.

15. True or False? Caffeine and epinephrine make the heart rate slow down.

16. The relaxation phase of the heart cycle is called _____________.

17. A defect in one or more heart valves can lead to a hissing heart sound called a(n) _____________.

18. Portable devices that analyze the heart's electric rhythm and apply electrical shocks as needed, known by the three-letter acronym _____________, are now available for use by the public to help someone having a heart attack.

Blood Vessels

19. Which one of the following statements is *false*?
    A. Veins, but not arteries, have one-way valves.
    B. Veins carry blood that is under very little blood pressure.
    C. Veins have the thinnest walls of any blood vessel.
    D. Arteries have layers of elastic connective tissue and smooth muscle.
    E. Blood pressure is the main force driving the blood from the heart to capillaries.

20. Blood is primarily moved through veins back to the heart by:
    A. the contraction of surrounding skeletal muscles and one-way valves in veins.
    B. smooth muscle contractions in the walls of capillaries and veins.
    C. the blood pressure created by heart contractions.
    D. the pull of gravity on the blood.
    E. the production of blood throughout the body.

21. True or False? Arteries carry blood <u>away from</u> the heart.

22. True or False? During strenuous exercise, blood is diverted <u>from</u> the digestive tract <u>to</u> skeletal muscles and the skin.

23. Persistent systolic blood pressure above 140 and diastolic pressure above 90 defines _____________.

Blood

24. Which one of the following statements about red blood cells is *false*? Red blood cells:
    A. are the most numerous type of blood cell.
    B. lack nuclei and other organelles.
    C. are well adapted to transport oxygen.
    D. have carbohydrates on their surface that determine the blood type.
    E. release clotting factors that convert fibrinogen into fibrin.

25. Which one of the following statements about platelets is *false*? Platelets:
    A. are the primary carriers of carbon dioxide in the blood.
    B. adhere to damaged tissue to form a sticky cluster that seals minor breaks.
    C. release clotting factors that convert fibrinogen into fibrin.
    D. help to plug leaks by forming clots.
    E. are bits of cytoplasm pinched off from larger cells in the bone marrow.

26. Compared to red blood cells, white blood cells:
    A. are smaller.
    B. have hemoglobin.
    C. have nuclei.
    D. are about 1000 times more abundant.

27. True or False? Leukemia is cancer of <u>erythrocytes</u>.

28. True or False? An abnormally low amount of hemoglobin or a low number of red blood cells is called <u>anemia</u>.

29. About half of the blood volume is a watery fluid called _____________.

30. Most oxygen in blood is bound to _____________ inside of red blood cells.

Cardiovascular Disease

31. Which one of the following is *not* a risk factor for developing cardiovascular disease?
    A. a diet high in cholesterol, trans fat, and saturated fat
    B. a diet high in fruits and vegetables
    C. smoking
    D. lack of exercise

32. True or False? The leading cause of death in the United States is <u>stroke</u>.

33. True or False? The complete blockage of a coronary artery will likely lead to a(n) <u>stroke</u>.

34. The circulatory system contributes to homeostasis by exchanging nutrients and wastes with _____________.

35. The chronic cardiovascular disease that results in a narrowing of blood vessels is called _____________.

Unifying Concepts of Animal Respiration

THE STRUCTURE AND FUNCTION OF RESPIRATORY SURFACES

36. Which one of the following respiratory systems uses a system of branching tubes to transport oxygen to nearly every cell in the body?
    A. lungs
    B. tracheal system
    C. gills
    D. body surface

37. True or False? In general, air holds <u>less</u> oxygen than water.

38. Insects breathe using _____________.

39. Extensions, or outfoldings, of the body surface typically used by aquatic organisms are called _____________.

40. Cellular respiration uses _____________ and glucose to produce energy-carrying _____________ molecules.

The Human Respiratory System

THE STRUCTURE AND FUNCTION OF THE HUMAN RESPIRATORY SYSTEM

41. As air passes into the respiratory system, it moves from:
    A. pharynx → larynx → trachea → bronchi → bronchioles → alveoli.
    B. pharynx → trachea → larynx → bronchioles → bronchi → alveoli.
    C. larynx → pharynx → trachea → bronchioles → bronchi → alveoli.
    D. pharynx → larynx → trachea → bronchioles → alveoli → bronchi.
    E. trachea → pharynx → larynx → bronchioles → alveoli → bronchi.

42. Which one of the following is *not* considered one of the three phases of gas exchange in humans?
    A. transport of O_2 from the extensively branched lungs to the rest of the body via the circulatory system
    B. production of ATP by aerobic metabolism consuming oxygen and producing carbon dioxide
    C. breathing
    D. the transport of O_2 from the extensively branched lungs to the rest of the body via the circulatory system

43. True or False? The inner surface of each <u>bronchus</u> is the respiratory surface.

44. Humans' vocal sounds are produced by flexing muscles in the voice box, which stretch the _____________ and make them vibrate.

TAKING A BREATH

45. Which of the following help us inhale?
    A. downward and inward movement of the ribs
    B. downward movement of the diaphragm
    C. increased pressure generated in the pharynx
    D. All of the above.
    E. None of the above.

46. True or False? Hiccups are caused by sudden contractions of the <u>lungs</u>.

47. Automatic control centers in the brain regulate breathing rate in response to levels of _____________ in the blood.

48. Expanding the chest cavity to inhale air is known as _____________ pressure breathing.

THE ROLE OF HEMOGLOBIN IN GAS TRANSPORT, HOW SMOKING AFFECTS THE LUNGS

49. Which one of the following statements about blood is *false*?
    A. Each molecule of hemoglobin consists of four polypeptide chains, each with a heme, at the center of which is an atom of iron.
    B. Most of the oxygen in blood is carried in hemoglobin molecules.
    C. Oxygen readily dissolves in blood.
    D. A shortage of iron causes a decrease in the rate of synthesis of hemoglobin.
    E. Iron deficiency is the most common cause of anemia.

50. Which one of the following statements about smoking is *false*?
    A. One of the worst sources of airborne pollutants is tobacco smoke.
    B. The carbon particles in cigarette smoke contain over 4,000 different chemicals.
    C. Tobacco smoke kills about 440,000 Americans every year.
    D. Secondhand cigarette smoke is not considered a health hazard.

51. True or False? More Americans die of <u>lung</u> cancer than any other form of cancer.

52. The most important lifestyle choice that impacts long-term health is _____________.

Evolution Connection: Choked Up

53. The first vertebrate lungs most likely appeared in:
    A. reptiles that had gills but could still walk on land.
    B. aquatic diving birds that also had gills.
    C. insects that could swim and fly.
    D. fish that had gills and lungs.
    E. amphibians already living on land.

54. True or False? The respiratory and digestive systems in humans share a common passageway in the back of the <u>trachea</u>.

55. The vertebrate lung system most likely evolved from the _____________ system.

Word Roots

alveoli = a cavity (alveoli: one of the dead-end, multilobed air sacs that constitute the gas exchange surface of the lungs)

atrio = a vestibule; **ventriculo** = ventricle (atrioventricular node: a region of specialized muscle tissue between the right atrium and right ventricle; it generates electrical impulses that primarily cause the ventricles to contract)

cardi = heart; **vascula** = a little vessel (cardiovascular system: the closed circulatory system characteristic of vertebrates)
erythro = red; **cyte** = cell (erythrocyte: a red blood cell)
fibrino = a fiber; **gen** = produce (fibrinogen: the inactive form of the plasma protein that is converted to the active form fibrin, which aggregates into threads that form the framework of a blood clot)
hyper = above, excessive; **tens** = stretched (hypertension: high blood pressure)
leuk = white; **emia** = blood (leukemia: cancer of leukocytes)
thrombo = a clot (thrombocytes: cells that promote blood clotting)

Key Terms

alveoli
anemia
arteries
arterioles
atherosclerosis
atrium
AV (atrioventricular) node
blood pressure
breathing
bronchi
bronchioles
capillaries
cardiac cycle
cardiovascular disease
cardiovascular system
circulatory system
closed circulatory system
coronary arteries
diaphragm
diastole
diffusion
double circulation system
erythrocytes
fibrin
fibrinogen
gills
heart
heart attack
heart murmur
heart rate
hemoglobin
hypertension
interstitial fluid
larynx
leukemia
leukocytes
lungs
negative pressure breathing
open circulatory system
pacemaker
pharynx
plasma
platelets
pulmonary circuit
pulse
red blood cells
respiratory surface
respiratory system
SA (sinoatrial) node
systemic circuit
systole
trachea
tracheal system
veins
ventricle
venules
vocal cords
white blood cells

Crossword Puzzle

Use the Key Terms list from this chapter to fill in the crossword puzzle.

ACROSS

1. the type of circuit that transports blood between the heart and lungs
3. the contraction stage of the heart cycle, when blood is actively pumped
5. the type of node that functions as the pacemaker of the heart
7. the rhythmic stretching that occurs when blood is forced into arteries
8. a specific type of surface where gases are exchanged between an animal and its environment
12. the plasma protein that is converted into fibrin
13. the SA node
14. a muscular organ that pumps blood
15. in animals, the vessels that return blood to the heart
18. the type of arteries that supply the heart
19. the sheet of muscle that expands the chest; it separates the chest cavity from the abdominal cavity
20. the typical respiratory surface of animals that live in water
21. the type of circulatory system in which blood is always contained in blood vessels or the heart
24. the death of heart muscle cells and the resulting failure of the heart
26. the type of system that transports gases and nutrients within the body
28. the stage of the heart cycle in which the heart muscle is relaxed
29. another name for high blood pressure
30. a type of system with a heart and branching network of blood vessels
32. the internal, saclike organ that exchanges gases with the environment
33. internal sacs deep within the lungs where gas exchange occurs
34. a threadlike protein that forms a dense network to create a patch where tissue is damaged
35. the type of cycle a heart goes through when it contracts rhythmically
37. the voice box, which contains the vocal cords
38. a pair of breathing tubes that branch off the trachea as they extend to the lungs
40. small arteries
42. the windpipe
44. a type of circulatory system in which blood is pumped through open-ended vessels and out among the body cells
45. the force that blood exerts against the walls of blood vessels
46. tiny air-conducting branches that extend from bronchi
48. the type of circulatory system that has two distinct circuits of blood flow
49. a heart chamber that receives blood from veins
51. a heart chamber that pumps blood out of a heart
52. the liquid matrix of the blood in which blood cells are suspended
54. the frequency of heart contractions
55. flexible muscles that vibrate in the larynx, producing vocal sounds

DOWN

1. the clotting elements in human blood, also called thrombocytes
2. the type of breathing in which air is pulled into the lungs
4. the type of circuit that transports blood between the heart and the rest of the body
6. an extensive network of branching internal tubes used by insects to breathe
9. the region in the digestive tract that receives food from the oral cavity
10. the type of system that functions in exchanging gases with the environment
11. another name for red blood cells
16. the type of disease that affects the heart and/or blood vessels
17. an iron-containing protein in red blood cells that binds to oxygen
22. the type of node that stimulates the ventricles to contract
23. an aqueous solution that surrounds body cells
25. another name for white blood cells
27. the movement of molecules down a concentration gradient
31. abnormally low amounts of hemoglobin or a shortage of red blood cells
33. vessels that carry blood away from the heart
36. a type of cardiovascular disease that causes blood vessels to narrow
37. cancer of white blood cells
39. the tiniest of all blood vessels
41. a defect in heart valves that results in a hissing sound
43. alternating processes of inhalation and exhalation
47. the type of blood cells that fight infections
50. small veins
53. the type of blood cells that transport oxygen

1
2
3
4
5
6
7
8
9
10
11
12
13
14
15
16
17
18
19
20
21
22
23
24
25
26
27
28
29
30
31
32
33
34
35
36
37
38
39
40
41
42
43
44
45
46
47
48
49
50
51
52
53
54
55

CHAPTER 24

The Body's Defenses

Studying Advice

a. Why do we need to get a flu shot every year? Why are some people allergic to certain substances? Why is AIDS so deadly? How does a vaccine prevent disease? Why are we so concerned about a bird flu pandemic? If these questions are ones you have wondered about in the past, this chapter might be of particular interest. The body's immune system is very complex and involves reactions that few students have ever considered. This chapter is your chance to get a peek inside this hidden and deadly world!

b. This chapter may be the most challenging chapter in the book. There are lots of cells and many interactions. The following organizing tables should help you keep track of what cells are doing and how they interact.

Student Media

Activities

Immune Responses

HIV Reproductive Cycle

BLAST Animations

Innate Immunity

Inflammation

Adaptive Immunity

MP3 Tutors

The Human Immune System

Process of Science

Why Do AIDS Rates Differ Across the U.S.?

What Causes Infections in AIDS Patients?

Videos

Discovery Channel Video: Fighting Cancer
Discovery Channel Video: Vaccines
T Cell Receptors

Organizing Tables

TABLE 24.1 Describe the development and functions of the following cells of the immune system.

Cell Type	How and Where It Develops	How It Functions
B cell		
T cell		
Effector cells		
Memory cells		
Helper T cells		
Cytotoxic T cells		

TABLE 24.2 Compare the following pairs, structures, or processes, noting similarities and important differences.

External defenses	Internal defenses
Antigen	Antibody
Humoral immune response	Cell-mediated immune response
Primary immune response	Secondary immune response
Autoimmune diseases	Immunodeficiency diseases

Content Quiz

Directions: Identify the *one* best answer for the multiple-choice questions. For true/false questions, determine if the statement is true or false. If false, change the underlined word(s) to make the statement true. Finally, add the correct word(s) to the fill-in-the-blank questions to make the statements true.

Biology and Society: An AIDS Vaccine Failure

1. Antibiotics are effective against ______________ infections.
   A. bacterial
   B. viral
   C. fungal
   D. All of the above.
   E. None of the above.

2. True or False? An effective vaccine against HIV has recently been developed.

3. True or False? HIV is deadly because it destroys the body's defenses against disease.

4. One way to prevent infections is to inject a harmless variant of a disease-causing microbe in the form of a ______________.

Innate Defenses

EXTERNAL INNATE DEFENSES

5. What happens to particles that are trapped in mucus lining the respiratory tract?
   A. Mucous cells digest the particles.
   B. Acids released by other cells lining the respiratory tract destroy the particles.
   C. The particles are surrounded by a tough protective capsule that prevents infection.
   D. The particles are moved upward until they are swallowed or expelled by coughing or sneezing.

6. True or False? Concentrated stomach acids kill most of the bacteria swallowed with food or saliva.

7. Lining the mucous membranes of the respiratory and digestive tracts, ______________ traps bacteria, dust, and other particles.

8. Sweat, saliva, and tears contain ______________, which disrupt the cell walls of bacteria.

INTERNAL INNATE DEFENSES

9. Which one of the following are white blood cells that release chemicals that kill diseased cells?
   A. neutrophils
   B. macrophages
   C. natural killer cells
   D. helper T cells

10. True or False? Interferon is produced by a virus-infected cell.

11. Some defensive proteins, called ______________ proteins, coat the surfaces of microbes, making them easier for macrophages to engulf.

12. Cells that engulf foreign cells, foreign molecules, and debris from dead cells by phagocytosis are called ______________ cells.

THE INFLAMMATORY RESPONSE

13. Which one of the following dampen(s) the normal inflammatory response and help(s) reduce swelling and fever?

    A. histamine

    B. prostaglandins

    C. interferons

    D. pyrogens

    E. ibuprofen

14. Which one of the following is a type of defense already present in the body without any requirement for preparation?

    A. innate defenses

    B. adaptive defenses

    C. routine defenses

    D. reactive defenses

15. True or False? Aspirin and ibuprofen treat the symptoms of an illness <u>and address</u> the underlying cause.

16. Chemicals that signal for an increase in blood flow to a damaged area are released by ______________ cells in the area.

THE LYMPHATIC SYSTEM

17. Which one of the following does *not* occur when your body is fighting an infection?

    A. Lymph nodes fill with a huge number of lymphocytes.

    B. Lymph can pick up microbes from infection sites just about anywhere in the body.

    C. Blood delivers microbes to the lymphatic nodes and organs.

    D. Phagocytic cells in lymphatic tissue engulf invaders.

18. True or False? The lymphatic vessels carry <u>blood</u>, which is similar to interstitial fluid.

19. The lymphatic system includes sac-like organs called ______________ that are packed with white blood cells.

Adaptive Defenses

20. Lymphocytes originate from stem cells in the:
    A. pancreas.
    B. spleen.
    C. thyroid.
    D. liver.
    E. bone marrow.

21. True or False? Exposure to specific pathogens elicits innate defenses.

22. True or False? Adaptive defenses depend upon white blood cells called erythrocytes.

23. The specific type of white blood cells associated with the thymus are ____________.

24. A foreign substance that elicits an immune response is a(n) ____________.

RECOGNIZING THE INVADERS

25. Which one of the following cell types secretes antibodies?
    A. T cells
    B. B cells
    C. helper T cells
    D. cytotoxic T cells
    E. None of the above.

26. Which one of the following statements about antibodies is *false*?
    A. At the tip of each arm of an antibody, a pair of variable regions forms an antigen-binding site.
    B. Each antibody molecule recognizes and binds to just one specific type of antigen.
    C. An antigen-binding site and the antigen it binds have complementary shapes.
    D. The structural variety of antibodies limits the immune system's ability to react to only a few kinds of antigen.

27. True or False? Monoclonal antibodies are produced by cells that are descended from a single cell.

28. True or False? Monoclonal antibody production produces antibodies that are specific for a single antibody.

29. True or False? We knew about the disease AIDS before we understood what pathogen caused it.

30. The overall shape of an antibody is most like the letter ____________.

31. Herceptin is a genetically engineered ____________.

32. AIDS is caused by the virus called ____________.

RESPONDING TO THE INVADERS

33. In the United States, widespread vaccination of children has virtually eliminated all of the following except:
   A. smallpox.
   B. polio.
   C. mumps.
   D. AIDS.

34. Which one of the following are involved in the cell-mediated immune response *and* the humoral immune response?
   A. T cells
   B. B cells
   C. memory cells
   D. effector cells
   E. None of the above.

35. Antibodies are carried to sites of infection in the body by:
   A. red blood cells.
   B. T cells.
   C. platelets.
   D. B cells.
   E. blood and lymph.

36. When an antigen activates only a tiny number of lymphocytes with receptors specific for a particular antigen, these selected cells proliferate in a process called:
   A. innate activation.
   B. sexual selection.
   C. clonal selection.
   D. the inflammatory response.
   E. vaccination.

37. Which one of the following statements about helper T cells is *false*?
   A. Helper T cells bind to other white blood cells that have previously encountered an antigen.
   B. Receptors embedded in the helper T cell's plasma membrane recognize and bind to the combination of a self protein and a nonself molecule displayed on a phagocytic cell.

C. An activated helper T cell grows and divides, producing more active helper T cells and memory cells.

D. An activated helper T cell inhibits the activity of cytotoxic T cells.

38. Which one of the following statements about cytotoxic T cells is *false*?

A. Cytotoxic T cells are the only cells in the body that produce antibodies.

B. Cytotoxic T cells bind to infected body cells.

C. An activated cytotoxic T cell might discharge the protein perforin, which makes holes in an infected cell's plasma membrane.

D. An activated cytotoxic T cell might discharge a protein that enters an infected cell and triggers a process that results in the death of the infected cell.

39. True or False? Antibodies produced against one particular antigen are usually <u>effective</u> against any other foreign substance.

40. True or False? The first response to exposure of lymphocytes to an antigen is called the <u>primary</u> immune response.

41. Effector T cells carry out ______________ immune responses.

42. The secretion of antibodies by effector B cells is part of the ______________ immune response.

43. A faster secondary immune response is produced when ______________ cells bind an antigen and multiply quickly, producing a large new clone of lymphocytes.

44. When the immune system detects an antigen, it produces defensive proteins called ______________.

45. In the process of ______________, the immune system is confronted with a harmless variant of a disease-causing microbe or one of its components.

Immune Disorders

ALLERGIES

46. Which one of the following does *not* occur during sensitization?

A. An allergen enters the bloodstream.

B. The allergen binds to B cells with complementary receptors.

C. B cells then proliferate through clonal selection.

D. B cells secrete large amounts of antibodies to that allergen.

E. B cells secrete large amounts of histamine, which triggers the inflammatory response.

47. True or False? Mast cells release <u>allergens</u>, which causes local inflammation that produces sneezing, coughing, and itching.

48. True or False? Allergic symptoms <u>are most common</u> in the nose and throat because allergens usually enter the body through these passageways.

49. Antigens that cause allergies are called ____________.

50. Some people are so extremely sensitive to certain allergens that any contact with them might cause a life-threatening response called ____________.

AUTOIMMUNE DISEASES

Match the autoimmune disease on the left to its description on the right.

_____ 51. rheumatoid arthritis

_____ 52. insulin-dependent diabetes

_____ 53. multiple sclerosis

_____ 54. lupus

A. antibodies are produced against many sorts of self molecules, including DNA

B. damage and painful inflammation of joints

C. destruction of myelin sheaths

D. destruction of cells of the pancreas

55. To prevent organ transplant rejection, doctors:
    A. employ cytotoxic T cells to destroy the B cells that will in turn destroy the transplant.
    B. promote anaphylactic shock in the organ recipient.
    C. look for a donor with self proteins matching the recipient's as closely as possible.
    D. first expose the person about to receive the transplant to antigens found on the organ to be transplanted.
    E. None of the above.

56. True or False? Every individual has a unique set of <u>self proteins</u>.

57. When the immune system turns against the body's own molecules, ____________ diseases occur.

IMMUNODEFICIENCY DISEASES, AIDS

58. Which one of the following statements about AIDS is *false*?
    A. HIV most often attacks helper memory cells.
    B. AIDS is an immunodeficiency disease.
    C. Most people with AIDS die from another infectious agent or from cancer.
    D. Education is the best weapon against the spread of AIDS.
    E. Since 1981, AIDS has killed more than 30 million people worldwide.

59. True or False? People with <u>SCID</u> are born with a marked deficit in T cells and B cells.

60. Hodgkin disease, radiation therapy, and drug treatments used against many cancers can all cause harm by affecting ______________, key cells of the immune system.

61. ______________ diseases occur in people who lack one or more of the components of the immune system.

62. HIV is deadly because it destroys the ______________ system, leaving the body defenseless against most invaders.

63. HIV most often attacks ______________ cells, the cells that activate other T cells and B cells.

Evolution Connection: HIV Evolution

64. The greatest obstacle to the eradication of AIDS is
    A. the ability of HIV to hide inside of cells.
    B. the inability to stop its transmission from person to person.
    C. the ability of HIV to trigger anaphylactic shock.
    D. the fast mutation rate of HIV.
    E. its ability to live for long periods outside of the body.

65. True or False? Drug-resistant strains of HIV <u>are not</u> found in newly infected patients.

66. HIV now rapidly adapts through the process of ______________ in a changing environment.

Word Roots

anti = against; **gen** = produce (antigen: a foreign macromolecule that does not belong to the host organism and that elicits an immune response)
auto = self (autoimmune diseases: when the immune system turns against the body's own molecules)
cyto = cell; **toxic** = poison (cytotoxic T cells: T cells that kill other cells)
humor = fluid (humoral immune response: the immunity conferred by B cells, which secrete antibodies into blood)
immuno = safe (immune system: the system that protects the body by recognizing and attacking pathogens and cancer cells)
lymph = fluid (lymph: a fluid, similar to interstitial fluid)
mono = one (monoclonal antibodies: the antibodies produced by a single clone of cells)

Key Terms

adaptive defenses
allergens
allergies
antibody
antigen
autoimmune diseases
B cells
cell-mediated immune response
clonal selection
complement proteins
cytotoxic T cells
effector cells
helper T cells
histamine
humoral immune response
immune system
immunodeficiency diseases
inflammatory response
innate defenses
interferons
lymph
lymph nodes
lymphatic system
lymphocytes
memory cells
monoclonal antibodies
pathogens
phagocytic cells
primary immune response
secondary immune response
T cells
vaccination
vaccine

Crossword Puzzle

Use the Key Terms list from this chapter to fill in the crossword puzzle.

ACROSS

2. an antigen that causes allergies
3. the type of T lymphocytes that kill infected and cancerous cells
5. the type of immunity that functions in defense against fungi, protists, bacteria, and viruses inside host cells
6. the type of defenses found only in vertebrates and must be activated by exposure to specific invaders
9. an abnormal sensitivity to an antigen in our surroundings
10. muscle cells or gland cells that performs the body's responses to stimuli
11. a substance released by injured cells that causes blood vessels to dilate during an inflammatory response
13. the colorless fluid, derived from interstitial fluid, in the lymphatic system of vertebrate animals
15. a foreign macromolecule that does not belong to the host organism and that elicits an immune response
17. the type of defenses that are already present in their final form and act the same whether or not an invader has been previously encountered
18. a harmless variant or derivative of a pathogen used to prevent disease
19. the system of vessels and lymph nodes, separate from the circulatory system
21. T lymphocytes that are used as a defense mechanism of the immune system
23. the type of response triggered by penetration of the skin or mucous membranes

27. the type of cells that remain in a lymph node until activated by the antigen that triggered its initial formation
28. the type of initial immune response to an antigen
29. a type of white blood cell that engulfs bacteria, foreign proteins, and the remains of dead body cells
30. a disease-causing virus or organism

DOWN

1. the mechanism that determines specificity and accounts for antigen memory in the immune system
3. a set of about 20 serum proteins that carries out a cascade of steps leading to the lysis of microbes
4. B lymphocytes that are used as a defense mechanism of our immune system
7. a type of immunological disease in which the immune system turns against itself
8. the type of T cell that is required by some B cells to help them make antibodies
9. an antigen-binding immunoglobulin produced by B cells
11. the type of immunity that uses antibodies to fight bacteria and viruses
12. the type of body system that protects the body by attacking foreign substances
14. the type of immune response elicited when an animal encounters an antigen again at some later time
16. a type of disease that decreases the body's ability to produce an effective immune response
20. a chemical messenger of the immune system that can help other cells resist a virus
22. organs located along lymph vessels that filter lymph
24. a white blood cell
25. the type of antibodies produced by cells descended from a single cell
26. a procedure that presents the immune system with a harmless variant or derivative of a pathogen

1
2
3
4
5
6
7
8
9
10
11
12
13
14
15
16
17
18
19
20
21
22
23
24
25
26
27
28
29
30

CHAPTER 25

Hormones

Studying Advice

a. This is a chapter that is very well suited for studying using note cards. The chapter introduces many glands and hormones with specific functions. A note-card type of studying system will help you master these fundamentals and organize the information in your mind.

b. As the authors note, textbook Table 25.1 summarizes the human endocrine system. It is a very useful table to refer to often. Mark this page with a piece of paper to make finding it easier as you read.

c. Complete the following table as you read to also help you remember the key functions of each gland. Check the accuracy of what you have written against the information in textbook Table 25.1.

Student Media

Activities

Overview of Cell Signaling
Peptide Hormone Action
Steroid Hormone Action
Human Endocrine Glands and Hormones

BioFlix

Homeostasis: Regulating Blood Sugar

BLAST Animation

Signaling: Endocrine
Signaling via Steroid Hormones

MP3 Tutor

Homeostasis and the Endocrine System

Process of Science

How Do Thyroxine and TSH Affect Metabolism?

Videos

Discovery Channel Video: Endocrine System

Organizing Table

TABLE 25.1 As you read, complete the table by describing the functions of each gland.

Gland	Functions
Hypothalamus	
Anterior pituitary	
Posterior pituitary	
Thyroid	
Parathyroid	
Islet cells of the pancreas	
Adrenal medulla	
Adrenal cortex	
Gonads	

Content Quiz

Directions: Identify the *one* best answer for the multiple-choice questions. For true/false questions, determine if the statement is true or false. If false, change the underlined word(s) to make the statement true. Finally, add the correct word(s) to the fill-in-the-blank questions to make the statements true.

Biology and Society: A Hormonal Fountain of Youth?

1. Hormone replacement therapy is used to relieve the decline of the hormone(s):
   A. estrogen.
   B. testosterone.
   C. prolactin.
   D. estrogen and progesterone.
   E. testosterone and prolactin.

2. True or False? Bioidenticals are synthetic human <u>hormones</u>.

3. Hormone replacement therapy is used to relieve the symptoms of ______________.

Hormones: An Overview

4. Which one of the following statements about how a steroid hormone works is *false*? A steroid hormone:
   A. enters a cell by diffusing through the plasma membrane.
   B. binds to a receptor protein in the cytoplasm or nucleus.
   C. triggers a signal-transduction pathway.
   D. and receptor complex attach to specific sites on the cell's DNA in the nucleus.

5. Which one of the following is *not* a type of hormone that binds to the plasma membrane of a target cell?
   A. amine hormones
   B. peptide hormones
   C. protein hormones
   D. nucleotide hormones

6. True or False? Hormones are primarily made and secreted by <u>exocrine</u> glands.

7. Derived from cholesterol, ______________ hormones bind to receptors inside a cell.

8. A regulatory chemical that travels from its production site through the blood to affect other sites in the body is a(n) ______________.

9. Cells that respond to a hormone are called ______________ cells.

10. The main body system for internal chemical regulation is the ______________ system.

The Human Endocrine System

11. Which one of the following has endocrine and nonendocrine functions?
    A. pancreas
    B. pituitary gland
    C. thyroid
    D. cartilage
    E. skeletal muscle

12. True or False? <u>Sex</u> hormones affect most of the tissues of the body.

13. Hormones from the pancreas help regulate levels of ______________ in the blood.

THE HYPOTHALAMUS AND PITUITARY GLAND

14. Which one of the following, if any, is *not* released by the pituitary gland?
    A. endorphins
    B. growth hormone
    C. antidiuretic hormone
    D. inhibiting hormones
    E. All of the above are released by the pituitary gland.

15. Which one of the following statements is *false*?
    A. The anterior pituitary is composed of nonnervous, glandular tissue.
    B. The posterior pituitary is composed of nervous tissue.
    C. The pituitary is the master control center of the entire endocrine system.
    D. Releasing hormones make the anterior pituitary secrete hormones.
    E. Inhibiting hormones make the anterior pituitary stop secreting hormones.

16. True or False? Alcohol <u>promotes</u> the release of antidiuretic hormone.

17. The anterior pituitary synthesizes and releases ______________ and ______________, which help regulate the menstrual cycle.

18. Serving as the body's painkillers, ______________ are made by the anterior pituitary gland.

19. Too little growth hormone during development can lead to ______________.

THE THYROID AND PARATHYROID GLANDS

20. Calcium in the body is needed to allow:
    A. muscles to function properly.
    B. nerve signals to be transmitted from cell to cell.
    C. cells to transport molecules across their membranes.
    D. blood to clot.
    E. All of the above.

21. Which of the following requires iodine to produce its hormones?
    A. adrenal medulla
    B. thyroid
    C. parathyroid
    D. pituitary
    E. All of the above.

22. Which of the following might result from *excessive* secretion of parathyroid hormone?
    A. fatal convulsions known as tetany
    B. gigantism
    C. hypothyroidism
    D. uncontrollable muscle contractions
    E. loss of calcium from bone

23. True or False? Hypothyroidism can result from dietary deficiencies or from a defective thyroid gland.

24. Many industrialized nations have reduced the frequency of ____________ by the incorporation of iodine into table salt.

25. Calcitonin and PTH are ____________ hormones because they have opposite effects.

26. The hormone ____________ causes calcium to be deposited in bones and makes the kidneys reabsorb less calcium as urine is formed.

THE PANCREAS

27. Which of the following best describes the relationship of insulin to glucagon?
    A. They work together to prepare the body to deal with stress.
    B. Insulin stimulates the pancreas to secrete glucagon.
    C. High levels of insulin inhibit pancreatic secretion of glucagon.
    D. They are antagonistic hormones.
    E. Insulin is a steroid hormone; glucagon is a protein hormone.

28. True or False? When the concentration of glucose in the blood rises, following the digestion of a meal, insulin is released.

29. A rise in the concentration of sugar in the blood is caused by the hormone ____________.

30. In the disease ____________, body cells are unable to absorb glucose from the blood.

31. Type I diabetes requires regular supplements of ____________.

32. Treatment for Type II diabetes includes exercise and controlling the intake of ____________ in the diet.

THE ADRENAL GLANDS

33. Which one of the following is *not* a response to epinephrine?
    A. release of glucose from the liver
    B. increased heart rate
    C. increased blood pressure
    D. increased absorption of glucose by the digestive tract
    E. increased breathing rate

34. Which one of the following glands is located nearest to the kidneys?
    A. pancreas
    B. pituitary gland
    C. parathyroid glands
    D. testes
    E. adrenal glands

35. True or False? Epinephrine and norepinephrine are secreted by the cells in the adrenal cortex.

36. Receiving ______________ for a long period makes a person more susceptible to infection.

The Gonads

37. Which, if any, of the following is *not* a category of sex hormone?
    A. prolactin
    B. estrogens
    C. androgens
    D. progestins
    E. All of the choices *are* sex hormones.

38. Which, if any, of the following would *not* be secreted in a sexually mature, healthy woman before she reaches menopause?
    A. testosterone
    B. estrogens
    C. progestins
    D. All of the choices *would* be secreted.

39. True or False? Females and males have all three types of sex hormones, but in different proportions.

40. The uterus is prepared to support a developing embryo by the hormones called ______________.

41. The female reproductive system is maintained by the hormones called ______________.

42. The development and maintenance of the male reproductive system is stimulated by the hormones called ____________.

Matching: Match the gland on the left to its best description on the right.

_____ 43. parathyroid

_____ 44. anterior pituitary

_____ 45. gonad

_____ 46. posterior pituitary

_____ 47. adrenal medulla

_____ 48. thyroid

_____ 49. islet cell

_____ 50. adrenal cortex

A. stores and releases antidiuretic hormone

B. releases PTH to help regulate calcium levels in blood

C. produces insulin and glucagon

D. produces glucocorticoids

E. produces hormones that regulate metabolic rate

F. synthesizes and releases FSH, LH, GH, and PRL

G. secretes androgens, estrogens, and progestins

H. secretes epinephrine and norepinephrine

THE PROCESS OF SCIENCE: CAN POLLUTION CAUSE SEX CHANGES?

51. Endocrine disruptors cause problems in aquatic animals because the disruptors:
    A. bind and stimulate the cellular protein receptors for androgens.
    B. destroy naturally occurring androgens and estrogens.
    C. prevent the body's tissues from responding to the signals of sex hormones.
    D. cause infertility in males by destroying the sperm-producing cells in the testes.

52. True or False? Endocrine disruptors in Florida were only found in water upstream from a paper mill.

53. Environmental pollutants that affect the endocrine system are called ____________.

Evolution Connection: Androgens, Anatomy, and Aggression

54. Which one of the following is *not* affected by androgens?
    A. male vocalizations
    B. regulation of blood calcium levels
    C. growth of facial hair
    D. aggressive behavior

55. True or False? Widespread sex determination by androgens suggests that this hormone took on this role late in evolution.

Word Roots

andro = male; **gen** = produce (androgens: the principal male steroid hormones, such as testosterone, which stimulate the development and maintenance of the male reproductive system and secondary sex characteristics)
endo = within (endorphins: the body's natural painkillers)
epi = above, over (epinephrine: a hormone produced as a response to stress; also called adrenaline)
gluco = sweet (glucagon: a hormone that increases blood sugar levels)
hyper = excessive (hyperthyroidism: the excessive production of thyroid hormones)
hypo = below (hypothalamus: the master control center of the endocrine system, located just below the thalamus in the brain)
mellit = honey (diabetes mellitus: a hormonal disease in which cells cannot obtain glucose from blood, resulting in high glucose levels in blood and urine)
para = beside (parathyroid glands: important endocrine glands located next to the thyroid gland)
pro = before; **lact** = milk (prolactin: a hormone produced by the anterior pituitary gland; it stimulates milk synthesis in mammals)

Key Terms

adrenal cortex
adrenal glands
adrenal medulla
androgens
antagonistic hormones
anterior pituitary
calcitonin
corticosteroids
diabetes mellitus
endocrine glands
endocrine system
endorphins
epinephrine
estrogens
glucagon
glucocorticoids
gonads
growth hormone (GH)
hormone
hypothalamus
insulin
norepinephrine
pancreas
parathyroid glands
parathyroid hormone (PTH)
pituitary gland
posterior pituitary
progestins
target cell
thyroid gland

Crossword Puzzle

Use the Key Terms list from this chapter to fill in the crossword puzzle.

ACROSS

2. an extension of the hypothalamus composed of nervous tissue that secretes hormones made in the hypothalamus
5. the types of cells that respond to a regulatory signal, such as a hormone
11. a hormone produced in response to stress
13. the internal system of the body that uses hormones for chemical communication
14. a hormone produced as a response to stress, also called adrenaline
17. a hormone produced in the brain and anterior pituitary that inhibits pain perception
20. the male or female sex organ
21. a mammalian thyroid hormone that lowers blood calcium levels
23. the abbreviation for the hormone that raises blood calcium levels
25. the type of endocrine gland located adjacent to the kidneys in mammals
26. the primary female steroid sex hormones
28. a vertebrate hormone that lowers blood glucose levels by promoting the uptake of glucose by most body cells
29. the endocrine gland at the base of the hypothalamus

DOWN

1. the endocrine glands that secrete a hormone that raises blood calcium levels
3. the type of hormone that counteracts the effects of another hormone
4. a gland that secretes digestive enzymes, insulin, and glucagon
6. a corticosteroid hormone secreted by the adrenal cortex
7. the endocrine gland that secretes iodine-containing hormones
8. a family of steroid hormones, including progesterone, produced by the mammalian ovary
9. an antagonistic hormone to insulin
10. any one of the many circulating chemical signals secreted by endocrine glands
12. a serious hormonal disease in which the body cells are unable to absorb glucose from the blood
13. the type of ductless glands that secrete hormones
15. part of a gland under the hypothalamus, it consists of endocrine cells that secrete hormones directly into blood
16. the outer portion of the adrenal gland
18. the family of steroids synthesized by and released from the adrenal cortex
19. the central portion of the adrenal gland
22. the principal male steroid hormones
24. the ventral part of the vertebrate forebrain that helps maintain homeostasis
27. the abbreviation for the hormone that stimulates development, growth, and metabolism

CHAPTER 26

Reproduction and Development

Studying Advice

a. Few systems of the body capture our interest and attention more than the ones we use to reproduce. This chapter is your chance to learn more about how our reproductive systems function. The topics range from the tragedy of sexually transmitted disease to the joy of having children.

b. This chapter is filled with details of the structures and functions of the reproductive system. Create a system of note cards or other quick question-and-answer review. Create your note cards as you read through the chapter. Then review them daily to master the long list.

Student Media

Activities

Reproductive System of the Human Male

Reproductive System of the Human Female

Sea Urchin Development

Frog Development

MP3 Tutor

Spermatogenesis and Oogenesis

The Female Reproductive Cycle

Embryonic Development

Process of Science

What Might Obstruct the Male Urethra?

What Determines Cell Differentiation in the Sea Urchin?

Videos

Hydra Budding
Hydra Releasing Sperm
Sea Urchin Embryonic Development (time-lapse)
Frog Embryo Development
Ultrasound of Human Fetus 1
Ultrasound of Human Fetus 2

Organizing Tables

TABLE 26.1 Compare the terms in the table. Provide at least one example of each.

	Definition	Example
Fission		
Fragmentation		
Regeneration		
Budding		

TABLE 26.2 Compare oogenesis and spermatogenesis in the table.

	Oogenesis	Spermatogenesis
Where each process occurs		
Number of gametes produced from each parent cell		
The distribution of cytoplasm in the cells produced by this process		
The relative size, motility, and nutrient storage in the gametes produced		
Timing of completion of the process		

TABLE 26.3 Identify the location where each of these processes typically occurs in humans.

	Location
Sperm storage	
Fertilization	
Implantation	

TABLE 26.4 Describe the structure and location of each of the developmental stages in the table.

	Structure	Location(s) in the Mother's Reproductive Tract
Cleavage		
Blastocyst		
Gastrula		

Content Quiz

Directions: Identify the *one* best answer for the multiple-choice questions. For true/false questions, determine if the statement is true or false. If false, change the underlined word(s) to make the statement true. Finally, add the correct word(s) to the fill-in-the-blank questions to make the statements true.

Biology and Society: High-Tech Babies

1. The recent increase in the number of multiple birth rates is most likely due to:
   A. better nutrition.
   B. reproductive technologies.
   C. an increase in the size of humans.
   D. increased use of vaccinations.

2. Newborns from multiple births are more likely to:
   A. be larger in size.
   B. be born after their expected due date.
   C. live longer and healthier lives.
   D. None of the above.

3. True or False? Because the chances of success using artificial fertilization are low, doctors often inject multiple embryos to increase the probability of a pregnancy.

4. One in seven American couples suffers from ____________, the inability to conceive after at least one year of trying.

Unifying Concepts of Animal Reproduction

ASEXUAL REPRODUCTION

Matching: Match the processes on the left to their best description on the right.

_____	5. reproduction	A.	the regrowth of lost body parts
_____	6. fission	B.	the creation of new individuals from existing ones
_____	7. regeneration	C.	splitting off new individuals from existing ones
_____	8. budding	D.	one individual splits into two or more about equal in size

9. Which one of the following statements is *false*? Compared to sexual reproduction, asexual reproduction:
   A. makes it easier to reproduce if an organism is sessile.
   B. makes it easier to reproduce if organisms are greatly isolated from one another.
   C. takes longer than sexual reproduction.
   D. allows an individual very well suited to its environment to quickly multiply and exploit available resources.
   E. eliminates the need to find a mate.

10. True or False? Asexual reproduction produces genetically diverse populations.

11. The creation of offspring that are genetically identical to a lone parent defines ____________ reproduction.

12. The splitting off of a new individual from existing ones is a form of asexual reproduction called ____________.

13. The regrowth of lost body parts is called ____________.

SEXUAL REPRODUCTION

14. Compared to asexual reproduction, sexual reproduction is generally more adaptive when:
   A. there is environmental stability.
   B. conditions favor the production of the greatest number of offspring.
   C. conditions favor the production of genetically identical offspring.
   D. the environment changes significantly.

15. True or False? Sexual reproduction creates genetically unique offspring through the blending of the genotypes of two parents.

16. True or False? The process of external fertilization occurs when sperm are deposited in or near the female reproductive tract and the gametes fuse within the female's body.

17. The male gamete is the ____________ and the female gamete is the ____________.

18. Two haploid sex cells unite during fertilization to form a diploid ____________.

19. Some species are ____________, possessing male and female reproductive systems.

Human Reproduction

MALE REPRODUCTIVE ANATOMY

Matching: Match the structures on the left to their best description on the right.

_____ 20. penis

_____ 21. testes

_____ 22. urethra

_____ 23. epididymis

_____ 24. scrotum

_____ 25. prostate

_____ 26. vas deferens

A. a coiled tube that stores sperm

B. one of the glands producing the fluid portion of semen

C. the site of gamete production in males

D. a sac containing the testes

E. a shaft that supports the glans

F. a tube that conducts sperm and urine

G. a duct sperm travel through during ejaculation

27. Which one of the following statements is *false*?
    A. About 95% of semen consists of sperm.
    B. The prostate gland is commonly diseased in men over 40.
    C. The testes function best at below-normal body temperature.
    D. The seminal vesicle adds part of the fluid that forms semen.
    E. Prostate cancer is the second most commonly diagnosed cancer in the United States.

28. True or False? The production of sperm is greatest at a temperature slightly above normal body temperature.

29. The expulsion of sperm-containing fluid from the penis is called ____________.

FEMALE REPRODUCTIVE ANATOMY

Matching: Match the structures on the left to their best description on the right.

_____ 30. cervix

_____ 31. follicle

_____ 32. oviduct

_____ 33. ovaries

_____ 34. vulva

_____ 35. uterus

_____ 36. vagina

_____ 37. clitoris

A. narrow neck at the bottom of the uterus

B. the site of gamete production in females

C. a short, sensitive shaft supporting a rounded glans

D. a single egg surrounded by one or more layers of cells

E. birth canal

F. common site of fertilization

G. the site of pregnancy

H. collective name for the outer female reproductive anatomy

38. Which one of the following statements is *false*?
    A. During a menstrual cycle, a woman typically ovulates about every 28 days.
    B. The uterus is about the size and shape of a pear.
    C. After about the ninth week of development, the embryo is called a fetus.
    D. During sexual arousal, the vagina, labia minora, and clitoris engorge with blood.
    E. The uterus is lined with a blood-rich layer of tissue called the myometrium.

39. True or False? It is recommended that women have their cervix examined yearly via a Pap test.

40. The organs that produce gametes are called ______________.

41. An egg cell is ejected from the follicle in the process of ______________.

GAMETOGENESIS

42. Which one of the following statements is *false*?
    A. Meiosis in oogenesis yields cells of unequal size.
    B. Meiosis in spermatogenesis yields cells of equal size.
    C. Meiosis I of spermatogenesis produces four secondary spermatocytes.
    D. Meiosis in human males occurs only within seminiferous tubules.
    E. Meiosis in females starts before birth.

43. Which one of the following statements is *false*?
   A. Polar bodies are produced during oogenesis but not spermatogenesis.
   B. Spermatogenesis produces four gametes but oogenesis results in only one gamete from each parent cell.
   C. Sperm are small, motile, and contain few nutrients, while eggs are large, nonmotile, and well stocked with nutrients.
   D. Human females create mature ova only during fetal development.
   E. Human males create new sperm every day from puberty through old age.

44. True or False? A secondary oocyte completes meiosis II after <u>fertilization</u>.

45. The creation of gametes within the gonads defines ____________.

46. At birth, a baby girl has thousands of ____________, diploid cells that have paused in prophase of meiosis I.

47. Sperm develop within ____________ in the testes.

48. Prior to ejaculation, sperm are stored within the ____________.

THE FEMALE REPRODUCTIVE CYCLE

49. Which one of the following statements is *false*?
   A. The menstrual discharge consists of blood, clusters of endometrial cells, and mucus.
   B. The first day of menstruation is designated as the middle of the menstrual cycle.
   C. The endometrium will not be discharged if an embryo implants.
   D. The menstrual discharge leaves the body through the vagina.
   E. After menstruation, the endometrium regrows, reaching its maximum thickness in 20 to 25 days.

50. True or False? The <u>menstrual</u> cycle controls the growth and release of an ovum.

51. True or False? The growth of an ovarian follicle is stimulated by <u>LH</u>.

52. Uterine bleeding caused by the breakdown of the endometrium defines ____________.

53. Ovulation typically takes place on day ____________ of the 28-day cycle.

54. FSH and LH are produced by the ____________.

55. A positive pregnancy test detects the hormone ____________.

Reproductive Health

CONTRACEPTION

Matching: Match the items on the left to their best description on the right.

_____ 56. tubal ligation

_____ 57. vasectomy

_____ 58. rhythm method

_____ 59. withdrawal

_____ 60. diaphragm

_____ 61. cervical cap

_____ 62. condom

_____ 63. spermicides

A. removing the penis from the vagina before ejaculation

B. removal of a short section from each vas deferens

C. a thimble-shaped, small cap that tightly covers the cervix

D. removal of a short section from each oviduct

E. sperm-killing chemicals

F. refraining from intercourse around the time of ovulation

G. a large, dome-shaped rubber cap that covers the cervix

H. a barrier that fits over the penis or within the vagina

64. Which one of the following statements is *false*?
    - A. Birth control pills typically contain synthetic estrogen and synthetic progesterone.
    - B. "The pill" prevents ovulation and keeps follicles from developing.
    - C. Combined hormone contraceptives are also available as a shot, a ring inserted into the vagina, or a patch.
    - D. Extensive evidence links the pill to cancers.
    - E. Birth control pills require a physician's examination and prescription.

65. True or False? There is <u>no chemical</u> contraceptive currently available that prevents the production or release of sperm.

66. If pregnancy has already occurred, the drug ______________ can be used to induce an abortion during the first seven weeks of pregnancy.

SEXUALLY TRANSMITTED DISEASES

67. Which one of the following statements is *false*?
    - A. AIDS is caused by HIV.
    - B. Very few STDs cause long-term problems or death if left untreated.
    - C. STDs are most prevalent among teenagers and young adults.
    - D. The best way to avoid unwanted pregnancy and the spread of STDs is abstinence.
    - E. Latex condoms provide the best dual protection for "safe sex."

68. True or False? Viral STDs are not curable.

69. True or False? Many people infected with sexually transmitted diseases have no apparent symptoms early in the infection.

70. Chlamydia, gonorrhea, and syphilis are caused by ____________.

71. Yeast infections are caused by ____________.

72. Genital herpes and genital warts are caused by ____________.

Human Development

FERTILIZATION

73. Which one of the following does *not* occur during fertilization?
    A. Fusion of egg and sperm changes the egg so that other sperm cannot penetrate it.
    B. Fructose from the semen fuels movement of the sperm's tail.
    C. Fusion of egg and sperm activates the egg's metabolic machinery.
    D. The sperm penetrates the jelly coat around the egg.
    E. Additional sperm surrounding the egg trigger the egg to undergo mitosis.

74. True or False? The sperm's thick head contains a diploid nucleus.

75. The enzymes that digest a hole in the jelly coat around the egg are found in a sperm's ____________.

76. The movement of the sperm tail is powered by ____________, organelles clustered near the middle of the sperm.

BASIC CONCEPTS OF EMBRYONIC DEVELOPMENT

Matching: Match each item on the left to its best description on the right.

_____ 77. endoderm
_____ 78. ectoderm
_____ 79. mesoderm
_____ 80. induction
_____ 81. blastocyst
_____ 82. gastrula
_____ 83. cleavage
_____ 84. programmed cell death

A. gives rise to the nervous system
B. results in an embryo shaped like a solid multicellular ball
C. gives rise to the digestive system
D. a hollow ball with three distinct tissue layers
E. a process used to carve out fingers and toes
F. gives rise to the heart, kidneys, and muscles
G. a way that cells influence an adjacent group of cells
H. a fluid-filled hollow ball of about 100 cells

85. Which one of the following does *not* occur during cleavage?

A. DNA replication

B. cytokinesis

C. the size of the embryo increases

D. mitosis

E. All of the above do occur during cleavage.

86. True or False? Changes in <u>cell shape</u> help to create embryonic structures.

PREGNANCY AND EARLY DEVELOPMENT

Matching: Match the structures on the left to their best description on the right.

_____ 87. yolk sac

_____ 88. allantois

_____ 89. amnion

_____ 90. chorion

A. it produces the embryo's first blood cells and its first gamete-forming cells in the gonads

B. a fluid-filled sac that encloses and protects the embryo

C. forms part of the umbilical cord

D. becomes part of the placenta

91. Which one of the following statements about the placenta is *false*?

A. The embryonic and maternal blood supplies in the placenta flow into each other.

B. Nutrients and oxygen are extracted from the mother's blood.

C. Embryonic wastes are released into the mother's blood.

D. Protective antibodies pass from the mother to the fetus.

E. Most drugs can cross the placenta and can harm the embryo.

92. True or False? Most viruses <u>cannot</u> cross the placenta and cause disease.

93. The outer cell layer of the gastrula, the _____________, becomes part of the placenta.

94. The _____________ is the organ that provides nourishment and oxygen to the embryo and helps dispose of its metabolic wastes.

95. The inner cell mass contains _____________ cells, which have the potential to give rise to every type of cell in the body.

THE STAGES OF PREGNANCY

96. Which one of the following does *not* occur during the third trimester?

A. The fetus gains the ability to maintain its own temperature.

B. The fetus's muscles thicken.

C. The fetus loses much of its fine body hair, except on its head.

D. The fetus rotates so that its head points upward toward the mother's lungs.

E. All of the above occur during the third trimester.

97. True or False? By the end of the <u>first</u> trimester, the fetus's eyes are open and its teeth are forming.

98. During the ____________ trimester, the fetus's circulatory system and respiratory system undergo changes that will permit the breathing of air.

99. All of a fetus's organs and major body parts are formed by the ____________ trimester.

100. The most dramatic changes in a fetus occur during the ____________ trimester.

CHILDBIRTH

101. Which one of the following statements about oxytocin is *false*? Oxytocin:
   A. is a powerful stimulant for the smooth muscles in the wall of the uterus.
   B. causes uterine contractions.
   C. promotes milk production by the mammary glands.
   D. is part of a positive-feedback control mechanism.

102. True or False? The pituitary hormone <u>estrogen</u> promotes milk production by the mammary glands.

103. The ____________ stage of labor is characterized by the birth of the child.

104. A child is born after a series of strong, rhythmic contractions of the uterus called ____________.

105. During the final stage of labor, the ____________ is delivered.

106. The longest stage of labor is ____________, lasting 6 to 12 hours or longer.

Reproductive Technologies

INFERTILITY

107. Infertility can be due to:
   A. low sperm counts.
   B. inability of sperm to swim to an egg.
   C. a lack of ova.
   D. failure to ovulate.
   E. All of the above.

108. Which one of the following statements is *false*?
   A. High sperm counts may be due to a scrotum kept too warm by tight-fitting underwear.
   B. Drug therapies (including Viagra) and penile implants can be used to treat impotence.
   C. Hormone injections can induce ovulation but may result in multiple pregnancies.
   D. As with sperm, ova can be obtained from donors willing to help infertile couples.
   E. A woman able to become pregnant but unable to support a growing fetus might hire a surrogate mother.

109. True or False? Infertility is usually due to problems with the <u>woman</u>.

110. Infertility can be caused by ____________, the inability to maintain an erection.

111. Temporary ____________ can result from alcohol or drug use, or because of psychological reasons.

IN VITRO FERTILIZATION

112. Which one of the following statements is *false*? *In vitro* fertilization:
   A. costs around $10,000 for each attempt, successful or not.
   B. begins with the surgical removal of ova and the collection of sperm.
   C. relies on the complete development of a fetus outside the uterus.
   D. permits genetic testing of embryos prior to implantation.

113. True or False? By implanting only embryos of a certain sex, *in vitro* fertilization <u>can be used</u> by couples to select the sex of their baby.

114. *In vitro* literally means ____________.

Evolution Connection: The "Grandmother Hypothesis"

115. Menopause may be adaptive because it:
   A. allows a woman to focus her energy on caring for the children she has.
   B. allows for a rest period before resuming reproduction.
   C. decreases the risks of cancer in the mother.
   D. increases milk production of the mother.

116. True or False? Most species <u>lose</u> their reproductive capacity throughout life.

117. The cessation of ovulation and menstruation is called ____________.

Word Roots

a = without (asexual reproduction: reproduction without sex)

blasto = produce; **cyst** = sac, bladder (blastocyst: a hollow ball of cells produced one week after fertilization in humans)

contra = against (contraception: the prevention of pregnancy)

ecto = outer; **derm** = tissue (ectoderm: the outermost tissue layer of an embryo)

ectomy = cut out (vasectomy: the cutting of each vas deferens to prevent sperm from entering the urethra)

endo = inner (endoderm: the innermost tissue layer of an embryo)

epi = above, over (epididymis: a coiled tubule located adjacent to the testes where sperm are stored)

fertil = fruitful (fertilization: the union of haploid gametes to produce a diploid zygote)

gastro = stomach, belly (gastrulation: the formation of a gastrula from a blastula)

in = without (infertility: the inability to reproduce sexually)

labi = lip; **major** = larger (labia majora: a pair of thick, fatty ridges that enclose and protect the labia minora and vestibule)

meso = middle (mesoderm: the middle tissue layer of an embryo)

oo = egg; **genesis** = producing (oogenesis: the process in the ovary that results in the production of female gametes)

ovi = egg (oviduct: the tube that carriers an egg from the ovary)

re = again (regeneration: the replacement of a body part)

tri = three (trimester: a three-month period)

tropho = nourish (trophoblast: the outer epithelium of the blastocyst, which forms the fetal part of the placenta)

Key Terms

allantois
amnion
asexual reproduction
birth control pills
blastocyst
budding
cervix
chlamydia
chorion
chorionic villi
cleavage
clitoris
condom
contraception
corpus luteum
diaphragm
ectoderm
egg
ejaculation
embryo
endoderm
endometrium
fertilization
fetus
fission
follicles
gametes
gametogenesis
gastrula
gastrulation
genital herpes
gestation
gonads
hermaphrodite
hymen
impotence
in vitro fertilization (IVF)
induction
infertility
inner cell mass
labia majora
labia minora
labor
menstrual cycle
menstruation
mesoderm
morning after pills (MAPs)
natural family planning
oogenesis
ovarian cycle
ovaries
oviduct
ovulation
ovum
oxytocin
penis
placenta
polar body
prepuce
primary oocyte
primary spermatocytes
programmed cell death
prostate gland
reproduction
reproductive cycle
rhythm method

(Continued)

scrotum	sexual reproduction	testes	vagina
secondary oocyte	sexually transmitted diseases (STDs)	trimesters	vas deferens
secondary spermatocytes	sperm	trophoblast	vasectomy
semen	spermatogenesis	tubal ligation	vulva
seminal vesicles	spermicides	umbilical cord	yolk sac
seminiferous tubules	stem cells	urethra	zygote
		uterus	

Crossword Puzzle

Use the Key Terms list from this chapter to fill in the crossword puzzle.

ACROSS

2. the outer epithelium of the blastocyst, which forms the fetal part of the placenta
5. in human development, one of three 3-month-long periods of pregnancy
7. a pouch of skin outside the abdomen that houses the testes
9. the lifeline between the embryo and the placenta
11. the cutting of each vas deferens to prevent sperm from entering the urethra
12. the type of oocyte that is a diploid cell, often arrested in prophase I of meiosis
14. the birth canal in mammals
15. the tube that transports sperm from the epididymis during ejaculation
16. the type of spermatocyte that undergoes meiosis I
17. the type of highly coiled tubules where sperm are produced
18. a gland that secretes a fluid component of semen that lubricates and nourishes sperm
19. the process of cytokinesis in animal cells
21. the hormone that induces contractions of uterine muscles and ejection of milk during nursing
22. the fluid that is ejaculated by a male during orgasm
23. a flexible sheath used to cover the penis during sexual intercourse
26. a type of haploid oocyte resulting from meiosis I in oogenesis, which will become an ovum after meiosis II
27. the prevention of pregnancy
28. a sensitive female sexual organ that becomes erect during sexual arousal
29. a dome-shaped rubber cap that covers the cervix
30. a recurring series of events that produce gametes in females
31. the three layered, cup-shaped embryonic stage in animals
32. a new developing individual
34. a common STD caused by a bacterial infection
35. the outer layer of three embryonic layers in a gastrula layers in animal embryos
38. the female gamete
39. another name for the rhythm method
40. chemical contraceptives taken orally that are very effective at preventing pregnancy
43. an extra-embryonic membrane that forms a fluid-filled embryonic sac
44. the inner lining of the uterus
46. the type of reproduction involving only one parent
49. a haploid cell such as an egg or sperm
50. the neck of the uterus that opens into the vagina
51. the production of eggs or sperm
54. a means of asexual reproduction by which one individual separates into two or more individuals of about equal size
55. a common STD caused by a virus

56. a developing human from the ninth week of gestation until birth
57. a fold of skin covering the head of the clitoris or penis
58. the male and female sex organs
59. a microscopic structure in an ovary that contains the developing ovum and secretes estrogens
63. the type of cycle in higher female primates that leads to a bloody discharge
65. the state of carrying developing young within the female reproductive tract
66. the production of female gametes
67. in mammals, the tube passing from the ovary to the uterus
68. a female organ in mammals where embryos develop
69. the abbreviation for a type of pill used as emergency contraception
70. the ability of one group of embryonic cells to influence the development of another
72. the release of an egg from ovaries
73. the inability to conceive a child
74. a tube that drains urine from the urinary bladder
75. rhythmic contractions of the uterus that give rise to the birth of a child
77. the collective name for the outer features of the female reproductive anatomy
78. the portion of a blastocyst that will form the baby
80. the middle of the three primary germ layers in animal embryos
81. the male reproductive gonad
82. a type of male gland that secretes an acid-neutralizing component of semen

DOWN

1. the process that produces sperm
3. a type of cell death
4. a means of sterilization in which a woman's oviducts are tied closed
6. sperm-killing chemicals
7. the abbreviation for contagious diseases spread by sexual contact
8. the cyclic recurrence of the follicular phase, ovulation, and the luteal phase in the mammalian ovary
10. a male gamete
12. a cell produced during meiosis that does not develop further or participate in fertilization
13. the diploid product of fertilization
17. a type of spermatocyte produced by meiosis I
18. the type of reproduction resulting from the joining of parental genetic material
19. the knobby outgrowths on the outside of the chorion
20. the innermost of the three primary germ layers in animal embryos
24. the creation of new individuals from existing ones
25. the remainder of a follicle after ovulation
30. a form of contraception also known as natural family planning
33. a means of asexual reproduction in which a new individual splits off after developing from an outgrowth of a parent
36. the expulsion of sperm-containing fluid from the penis
37. an extra-embryonic membrane that stores embryonic nitrogenous waste
40. the embryonic stage consisting of a hollow ball of cells

DOWN

41. a thin membrane that partly covers the vaginal opening in a human female
42. the union of haploid gametes to produce a diploid zygote
45. the extraembryonic membrane that contributes to the formation of the mammalian placenta
47. a female gamete
48. the process that forms a gastrula
52. an extra-embryonic membrane that stores yolk in bird and reptile eggs
53. a structure in the pregnant uterus for nourishing a viviparous fetus with the mother's blood supply
60. the inability to maintain an erection
61. a pair of thick, fatty ridges that enclose and protect the labia minora and vestibule
62. an individual that produces sperm and eggs
63. the shedding of portions of the endometrium during a menstrual cycle
64. a pair of thin ridges that enclose the vestibule
66. the female reproductive gonad
71. the abbreviation for the process of fertilizing ova in laboratory containers
76. the copulatory structure of male mammals
79. a type of cell that gives rise to other cells and restores its own population

CHAPTER 27

Nervous, Sensory, and Motor Systems

Studying Advice

a. Have you ever thought about how you think? Ever wondered how your brain keeps everything sorted out? Ever consider how you remember anything? This fascinating chapter reveals insights into many of these amazing processes.

b. This is not a chapter for one night of studying. It is long and contains an extensive list of vocabulary terms. The following organizing tables help you sort out some of the information, but you will need to create other systems to learn the vocabulary. Consider using the list of key terms toward the end of this chapter to create note cards, with the terms on one side and the definitions on the other. You will need to quiz yourself over many days to master all of these terms.

Student Media

Activities

Neuron Structure

Nerve Signals: Action Potentials

Signal Transmission at a Chemical Synapse

Structure and Function of the Eye

Human Skeleton

Skeletal Muscle Structure

Muscle Contraction

BioFlix

How Neurons Work

How Synapses Work

Muscle Contraction

BLAST Animations

Action Potential
Signal Transmission at Synapses
Signal Amplification in Neurons
Anatomy of Muscle

MP3 Tutors

Neurons and Electrical Potentials
The Human Brain
Sensory Receptors
Muscle

Process of Science

What Triggers Nerve Impulses?
How Do Electrical Stimuli Affect Muscle Contraction?

Videos

Discovery Channel Video: Teen Brains
Discovery Channel Video: Muscles and Bones

Organizing Tables

TABLE 27.1 **Compare the two main subdivisions of the nervous system in the table. In the function(s) category, indicate which systems are responsible for sensory input, integration, and motor output.**

	Structural Components	Function(s)
CNS		
PNS		

TABLE 27.2 **Compare the structures and functions of the parts of a neuron in the table.**

	Structural	Function(s)
Cell body		
Dendrite		
Axon		

TABLE 27.3 **Compare the structures and functions of the components of the peripheral nervous system in the table.**

	Structural Components	Function(s)
Sensory division		
Motor division: Somatic nervous system		
Motor division: Autonomic nervous system, sympathetic division		
Motor division: Autonomic nervous system, parasympathetic division		

TABLE 27.4 **Compare the five general categories of sensory receptors in the table.**

	Location	Function(s)
Pain receptors		
Thermoreceptors		
Mechanoreceptors		
Chemoreceptors		
Electromagnetic receptors		

Content Quiz

Directions: Identify the *one* best answer for the multiple-choice questions. For true/false questions, determine if the statement is true or false. If false, change the underlined word(s) to make the statement true. Finally, add the correct word(s) to the fill-in-the-blank questions to make the statements true.

Biology and Society: Beyond Human Experience

1. Which of the following systems are found in some vertebrates but not humans?
   A. magnetoreception
   B. electroreception
   C. echolocation
   D. All three of the above are not found in humans.
   E. Humans have all three systems.

2. True or False? By detecting weak electrical fields using <u>magnetoreception</u>, hammerhead sharks can find prey buried under sand at the bottom of the ocean.

3. Bats and dolphins can detect the echoes of emitted sound waves to better understand their environment in a process called _____________.

An Overview of Animal Nervous Systems

ORGANIZATION OF NERVOUS SYSTEMS, NEURONS

4. If you touch something sharp and pull your finger back to avoid injury, the signals travel from:
   A. sensory input to integration to motor output.
   B. sensory input to motor output to integration.
   C. integration to sensory input to motor output.
   D. motor output to integration to sensory input.
   E. motor output to sensory input to integration.

5. True or False? Signals from the central nervous system are sent out to effectors by <u>sensory</u> neurons.

6. True or False? The <u>central</u> nervous system is mostly composed of nerves that carry signals into and out of the CNS.

7. A(n) _____________ is a communication line made from cable-like bundles of neuron fibers tightly wrapped in connective tissue.

8. The conveyance of signals to the CNS from sensory receptors is called _____________.

9. The part of a neuron that houses the nucleus is called the _____________.

10. Outnumbering neurons are _____________ cells that protect, insulate, and reinforce the neurons.

11. Axons that convey signals very rapidly are enclosed along most of their length by an insulating material called the _____________.

12. Signals are carried toward another neuron or toward an effector by _____________.

13. The eyes are an example of _____________, the places where sensory signals are generated.

SENDING A SIGNAL THROUGH A NEURON

14. Stimulating a neuron's plasma membrane to generate a nerve signal is most like:
    A. hitting a baseball with a bat.
    B. turning on a flashlight to create light.
    C. bouncing a ball against the ground.
    D. cutting up paper with scissors.
    E. constructing a dog house.

15. Which one of the following statements about membrane potentials is *false*?
    A. The potential energy of a neuron is in an electrical charge difference across a neuron's plasma membrane.
    B. During a resting membrane potential, the cytoplasm just inside the membrane is negative in charge.
    C. A membrane stores energy by joining opposite charges together.
    D. A cell's membrane has channels and pumps that regulate the passage of positive ions, contributing to the resting membrane potential.
    E. During a resting potential, the fluid just outside the cell is positive.

16. How do the events of an action potential cause the "domino effect" of a nerve signal?
    A. No ions are allowed to pass through the membrane where an action potential occurs, so ions cross at the next available point.
    B. Inflowing negative ions trigger the opening of channels in the membrane next to the action potential.
    C. Inflowing positive ions trigger the opening of channels in the membrane next to the action potential.
    D. Changes in the myelin sheath create new openings in regions next to the place where an action potential is occurring.

17. How do action potentials relay different intensities of information to the central nervous system?
    A. A stronger action potential is sent when a signal is stronger.
    B. A weaker action potential is sent when a signal is stronger.
    C. Stronger signals cause a greater surge of ions across the membrane.
    D. The frequency of action potentials changes with the intensity of stimuli.

18. True or False? Action potentials are <u>all-or-none</u> events.

19. True or False? The cytoplasm just inside a neuron is <u>positively</u> charged.

20. The voltage across a plasma membrane of a resting neuron is called the _____________.

21. Any factor that causes a nerve signal to be generated is called a(n) _____________.

22. If a stimulus is strong enough, a sufficient number of channels open to reach the _____________, the minimum change in a membrane's voltage that must occur to trigger an action potential.

PASSING A SIGNAL FROM A NEURON TO A RECEIVING CELL

Matching: Match each item on the left to its best description on the right.

_____ 23. caffeine

_____ 24. nicotine

_____ 25. alcohol

_____ 26. opiates

_____ 27. drugs to treat schizophrenia

_____ 28. Ritalin

_____ 29. endorphins

_____ 30. Parkinson's disease

_____ 31. schizophrenia

A. naturally decrease our perception of pain

B. chemically similar to dopamine and norepinephrine

C. a strong depressant

D. a disease associated with a lack of dopamine

E. binds to and activates receptors for acetylcholine

F. counters the effects of inhibitory neurotransmitters

G. a disease associated with an excess of dopamine

H. drugs that bind to endorphin receptors

I. block dopamine receptors

32. Where are neurotransmitters located before an action potential travels down a nerve?
    A. in the synaptic cleft
    B. in the synaptic terminal
    C. in the membrane of the neuron
    D. in the dendrites
    E. in the nerve cell body

33. What happens to neurotransmitters after they bind to a receiving neuron's plasma membrane?
    A. They are absorbed by the receiving neuron's plasma membrane.
    B. They are sent to the nerve cell body.
    C. They travel down the receiving neuron's plasma membrane.
    D. They are broken down or transported back to the sending neuron for recycling.

34. True or False? Synapses <u>can be</u> either electrical or chemical.

35. True or False? In <u>a chemical</u> synapse, an action potential jumps directly from one neuron to the next.

36. The narrow gap separating a synaptic knob of the sending neuron from the receiving neuron is called a(n) _____________.

37. A relay point between two neurons or between a neuron and an effector cell is a(n) _____________.

38. In a(n) _____________ synapse, an action potential is converted to chemical signals.

39. A(n) _____________ carries information from one nerve cell to another in a chemical synapse.

The Human Nervous System: A Closer Look

THE CENTRAL NERVOUS SYSTEM

40. Which one of the following statements about the spinal cord is *false*?
    A. The spinal cord functions like a telephone cable jam-packed with wires.
    B. The spinal cord is protected by the spinal column.
    C. Meninges surround the brain and spinal cord.
    D. The center of the spinal cord is solid.

41. True or False? All vertebrates have a central nervous system distinct from a peripheral nervous system.

42. True or False? The peripheral nervous system integrates information coming from the senses.

43. True or False? Paralysis of the lower half of the body is called quadriplegia.

44. The most serious type of meningitis is usually _____________ meningitis.

45. The _____________ is the master control center of the nervous system.

46. An infection of the meninges is called _____________.

THE PERIPHERAL NERVOUS SYSTEM

47. Which one of the following does *not* occur when the sympathetic division of the autonomic nervous system is activated?
    A. bronchi dilate
    B. heart rate accelerates
    C. epinephrine and norepinephrine are released into the bloodstream
    D. breathing rate decreases

48. True or False? The autonomic nervous system is said to be voluntary, because most of its actions are under conscious control.

49. True or False? The sympathetic division primes the body for digesting food and resting.

50. True or False? Neurons of the autonomic nervous system carry signals to skeletal muscles.

51. If you suddenly realize that you missed a lecture exam, the _____________ division of your autonomic nervous system will likely respond.

THE HUMAN BRAIN

Matching: Match each structure on the left to its best description on the right.

_____ 52. thalamus
_____ 53. hypothalamus
_____ 54. brainstem
_____ 55. cerebellum
_____ 56. corpus callosum
_____ 57. cerebral cortex
_____ 58. bipolar disorder

A. the highly folded outer surface of the cerebrum
B. consists of the medulla oblongata, pons, and midbrain
C. regulates body temperature, blood pressure, hunger, and thirst
D. involves extreme mood swings from manic to depressed phases
E. provides coordination of movement and balance
F. a band that connects the cerebral hemispheres
G. sorts data into categories; suppresses or enhances other signals

59. Which one of the following statements is *false*?
   A. The cerebral cortex is divided into right and left sides.
   B. The corpus callosum restricts communication between the two hemispheres.
   C. Each side of the cerebral cortex has four lobes.
   D. The association areas are the sites of higher mental activities.
   E. Language results from extremely complex interactions among several association areas.

60. True or False? The <u>medulla oblongata</u> is a planning center for body movements.

61. Areas in the two cerebral hemispheres become specialized for different functions in the process of _____________.

62. A form of mental deterioration or dementia, _____________ disease is a progressive illness seen in about 35% of people at age 85.

The Senses

SENSORY INPUT

Matching: Match each structure on the left to its best description on the right.

_____ 63. pain receptors
_____ 64. thermoreceptors
_____ 65. mechanoreceptors
_____ 66. chemoreceptors
_____ 67. electromagnetic receptors

A. detect various forms of mechanical energy
B. respond to chemicals
C. respond to energy of various wavelengths
D. detect either heat or cold
E. respond to excess heat or pressure

68. Which of the following receptors, if any, are *not* found in human skin?
   A. thermoreceptors
   B. electromagnetic receptors
   C. pain receptors
   D. mechanoreceptors
   E. All of the above are found in the skin.

69. True or False? Receptor cells convert one type of signal (the stimulus) into an electrical signal in a process called sensory <u>adaptation</u>.

70. True or False? The stronger the stimulus, the <u>stronger</u> the receptor potential.

71. The change in membrane potential in a receptor cell is called the ______________.

72. Local regulators that increase pain by sensitizing pain receptors are called ______________.

VISION

Matching: Match each structure on the left to its best description on the right.

_____ 73. astigmatism
_____ 74. farsightedness
_____ 75. vitreous humor
_____ 76. nearsightedness
_____ 77. cones
_____ 78. rods
_____ 79. tear gland
_____ 80. conjunctiva
_____ 81. glaucoma
_____ 82. conjunctivitis
_____ 83. retina
_____ 84. choroid
_____ 85. iris
_____ 86. pupil
_____ 87. sclera
_____ 88. cornea

A. an inflammation of the conjunctiva
B. a pigmented layer between the sclera and retina
C. a mucous membrane that helps keep the eye moist
D. a tough, whitish layer of connective tissue
E. secretes a dilute salt solution
F. blurred vision caused by a misshapen lens or cornea
G. the opening in the center of the iris
H. it lets light into the eye and focuses light onto the lens
I. a layer that contains photoreceptor cells
J. jellylike material filling the large chamber behind the lens
K. it gives the eye its color
L. detect shades of gray in the retina
M. ability to see far but not near
N. caused by increased pressure inside the eye
O. ability to see near but not far
P. detect color in the retina

89. Which one of the following statements is *false*?

A. The lens of the eye focuses light onto the retina.

B. The shape of the lens is controlled by muscles attached to the choroid.

C. When the eye focuses on a nearby object the lens becomes thinner and flatter.

D. There are no photoreceptor cells in the part of the retina where the optic nerve passes through the back of the eye.

E. An infection or allergic reaction may cause inflammation of the conjunctiva, a condition called conjunctivitis.

90. True or False? The much smaller chamber in front of the lens contains a thin fluid, the <u>vitreous</u> humor.

91. Rods contain a visual pigment called _____________, which can absorb dim light. Cones contain a visual pigment called _____________.

HEARING

Matching: Match each structure on the left to its best description on the right.

_____ 92. pinna

_____ 93. auditory canal

_____ 94. Eustachian tube

_____ 95. cochlea

_____ 96. organ of Corti

_____ 97. eardrum

A. a tube extending from the pinna to the eardrum

B. the bendable structure we commonly call our "ear"

C. a long coiled tube in the inner ear

D. an array of hair cells embedded in a basilar membrane

E. conducts air between the middle ear and the back of the throat

F. a sheet of tissue that separates the outer ear from the middle ear

Sequencing: For questions 98–106, indicate the correct sequence of the following nine events that occur in the process of hearing. Place a 1 in front of the first event, a 2 in front of the second, etc.

_____ 98. Hair cells develop a receptor potential and release more neurotransmitter molecules at its synapse with a sensory neuron.

_____ 99. As a pressure wave passes through the cochlea, it pushes downward and makes the basilar membrane vibrate.

_____ 100. When a hair cell's projections are bent, ion channels in its plasma membrane open, and positive ions enter the cell.

_____ 101. A vibrating object creates pressure waves in the surrounding air.

_____ 102. Vibrations pass through the hammer, anvil, and stirrup in the middle ear.

_____ 103. The sensory neuron sends more action potentials to the brain through the auditory nerve.

_____ 104. Waves make the eardrum vibrate with the same frequency as the sound.

_____ 105. Vibration of the basilar membrane makes the hairlike projections on the hair cells alternately brush against and draw away from the overlying membrane.

_____ 106. The stirrup transmits the vibrations to the inner ear, producing pressure waves in the fluid within the cochlea.

107. The actual sensory transduction converting sound into an action potential occurs in the:
   A. outer ear.
   B. middle ear.
   C. inner ear.
   D. All of the above.
   E. None of the above.

108. Deafness can be caused by:
   A. the inability to conduct sounds.
   B. a ruptured eardrum.
   C. stiffening of the middle-ear bones.
   D. damage to receptor cells or neurons.
   E. All of the above.

109. True or False? The three small bones are located in the <u>inner</u> ear.

110. A ringing or buzzing sound in your ears is a condition called _____________.

Motor Systems

THE SKELETAL SYSTEM

111. Which one of the following statements about osteoporosis is *false*?
   A. Prevention of osteoporosis begins with sufficient calcium intake while bones are still increasing in density.
   B. Walking, jogging, and lifting weights builds bone mass and is beneficial throughout life.
   C. Increased levels of estrogen contribute to osteoporosis.
   D. Insufficient exercise, an inadequate intake of protein and calcium, smoking, and diabetes mellitus may contribute to osteoporosis.
   E. Treatments include calcium and vitamin supplements, hormone replacement therapy, and drugs that slow bone loss or increase bone formation.

112. Which one of the following is an autoimmune disease in which the joints become highly inflamed and their tissues possibly destroyed?
   A. rheumatoid arthritis
   B. Parkinson disease
   C. tinnitus
   D. osteoporosis
   E. multiple sclerosis

113. True or False? The <u>axial</u> skeleton is made up of the bones of the limbs, shoulders, and pelvis.

114. True or False? Humans, like all vertebrates, have an <u>endoskeleton</u>, hard supporting elements situated among soft tissues.

115. True or False? The shaft of a long bone surrounds a central cavity that contains <u>red</u> bone marrow, mostly stored fat brought into the bone by the blood.

116. A(n) _____________ joint allows us to rotate the forearm at the elbow.

117. A freely moving _____________ joint joins the humerus to the scapula and the femur to the pelvis.

118. A(n) _____________ joint between the humerus and the head of the ulna permits movement in a single plane.

THE MUSCULAR SYSTEM, STIMULUS AND RESPONSE: PUTTING IT ALL TOGETHER

119. Which one of the following does *not* occur during a muscle contraction?
   A. A myosin head gains energy from the breakdown of NADH.
   B. An energized myosin head binds to an exposed binding site on actin.
   C. The molecular event that actually causes sliding is called the power stroke.
   D. During the power stroke, energy is released from the myosin head, and the head bends back to its low-energy position, pulling the thin filament toward the center of the sarcomere.
   E. On the next power stroke, the myosin head attaches to another binding site ahead of the previous one on the thin filament.

120. Which one of the following summarizes the activities of actin and myosin during a muscle contraction?
   A. cock, detach, attach, bend
   B. detach, cock, attach, bend
   C. bend, cock, detach, attach
   D. attach, cock, bend, detach
   E. bend, detach, attach, cock

121. A motor unit functions most like:
   A. blowing into a tuba to make music.
   B. breaking a stick in half.
   C. burning gasoline to make a truck engine function.
   D. a switch that controls a set of lights.
   E. assembling pieces of a puzzle.

122. The arrangement of sarcomeres in a myofibril is most like:
   A. railroad cars in a train.
   B. checkers on a checkerboard.
   C. leaves on a tree.
   D. clouds in the sky.

123. Which one of the following is an example of a motor response by a baseball player?
   A. seeing a ball
   B. hearing the crack of the bat against the ball
   C. determining where the ball will go
   D. reaching out to catch the ball
   E. feeling joy in success

124. True or False? The shortening of <u>sarcomeres</u> causes muscles to shorten.

125. True or False? <u>Thin</u> filaments are composed of myosin.

126. True or False? A sarcomere contracts when its thin filaments <u>slide across</u> its thick filaments.

127. Muscles attach to bone by ______________.

128. A myofibril consists of repeating units called ______________.

129. Each muscle fiber consists of a bundle of smaller ______________.

130. A motor neuron and all the muscle fibers it controls defines a(n)______________.

131. A motor neuron forms synapses with the muscle fibers at ______________ junctions.

Evolution Connection: Seeing UV

132. The ability of birds to see UV light appears to have evolved by:
   A. the duplication of a gene.
   B. the duplication of a chromosome.
   C. the fusion of two chromosomes.
   D. a single amino acid change in the pigment protein rhodopsin.

133. True or False? Electroreception in some fish resulted from a modification of <u>muscle</u> cells.

Word Roots

aqua = water; **humor** = fluid (aqueous humor: a thin, watery fluid in front of the lens of the eye)
arthro = joint; **itis** = inflammation (arthritis: inflammation of the joints of the skeleton)
audi = hear (auditory canal: the tube that channels sound to the eardrum)
auto = self (autonomic nervous system: a subdivision of the motor nervous system of vertebrates that regulates the internal environment)
bi = two (bipolar disorder: a manic-depressive disorder in which a person switches between two extremes)
cerebro = brain (cerebrum: the largest and most sophisticated part of the human brain)
chemo = chemical (chemoreceptors: sensory cells that respond to chemicals)
dendro = tree (dendrite: one of usually numerous, short, highly branched processes of a neuron that conveys nerve impulses toward the cell body)
endo = inner (endoskeleton: a skeleton inside the body)
hypo = below (hypothalamus: the master control center of the endocrine system, located below the thalamus in the brain)
inter = between (interneurons: an association neuron; a nerve cell within the central nervous system that forms synapses with sensory and motor neurons and integrates sensory input and motor output)
myo = muscle (myofibril: a contractile thread in a muscle cell)
neuro = nerve; **trans** = across (neurotransmitter: a chemical messenger released from the synaptic terminal of a neuron at a chemical synapse that diffuses across the synaptic cleft and binds to and stimulates the postsynaptic cell)
osteo = bone; **pori** = a hole (osteoporosis: a disease of thinning and porous bones)
para = near (parasympathetic division: one of two divisions of the autonomic nervous system)
photo = light (photopsin: a light-sensing pigment found in cones)
soma = body (somatic nervous system: a branch of the nervous system that carries signals to skeletal muscles)
syn = together (synapse: the locus where a neuron communicates with a postsynaptic cell in a neural pathway)
thermo = heat (thermoreceptors: sensory receptors that detect heat)

Key Terms

action potential
Alzheimer's disease
appendicular skeleton
aqueous humor
arthritis
association areas
astigmatism
auditory canal
autonomic nervous system
axial skeleton
axon
ball-and-socket joints
bipolar disorder
brain
brainstem
cell body
central nervous system (CNS)
cerebellum
cerebral cortex
cerebrospinal fluid
cerebrum
chemoreceptors
cochlea
cones
cornea
corpus callosum
dendrites

(Continued)

eardrum
effectors
electromagnetic receptors
endoskeleton
Eustachian tube
farsightedness
hinge joint
hypothalamus
inner ear
integration
interneurons
iris
lateralization
lens
ligaments
major depression
mechanoreceptors
medulla oblongata
meninges
middle ear
motor neurons
motor output
motor units
muscular system
myelin sheath
myofibrils
nearsightedness
nerve
nervous system
neurons
neurotransmitter
optic nerve
osteoporosis
outer ear
pain receptors
parasympathetic division
peripheral nervous system (PNS)
photoreceptors
pinna
pivot joint
pons
pupil
receptor potential
resting potential
retina
rheumatoid arthritis
rods
sarcomeres
sclera
sensory input
sensory neurons
sensory transduction
skeletal muscle
skeletal system
somatic nervous system
spinal cord
stimulus
supporting cells
sympathetic division
synapse
synaptic cleft
synaptic terminal
tendons
thalamus
thermoreceptors
thick filaments
thin filaments
threshold
vitreous humor

Crossword Puzzle

Use the Key Terms list from this chapter to fill in the crossword puzzle.

ACROSS

2. the complex, coiled organ of hearing that contains the organ of Corti
3. layers of connective tissue between the brain and spinal cord
4. a sensory structure that responds to chemical changes
6. the sites of higher mental activities
8. blurred vision caused by a misshapen lens or cornea
9. the transparent frontal portion of the sclera, which admits light into the vertebrate eye
12. plasmalike liquid in the space between the lens and the cornea
13. the part of the vertebrate hindbrain that coordinates movement and balance
14. a relay point at the tip of a transmitting neuron's axon
15. the colored portion of the choroid at the front of the eye
18. inflammation of the joints
19. the type of joint that enables mammals to move arms and legs in several planes
22. the part of the skeleton associated with the limbs and girdles
23. a ropelike bundle of neuron fibers tightly wrapped in connective tissue
24. the tube that channels sound waves to the eardrum
25. the disease that is a form of dementia
27. the type of nerve that carries sensory signals from the eye
31. the interpretation of sensory signals and the formulation of responses within the central nervous system
32. blood-derived fluid that surrounds and cushions the brain and spinal cord
36. a type of disorder that is also called manic-depressive disorder
38. the region of the vertebrate ear that includes the cochlea, organ of Corti, and semicircular canals
39. the flaplike part of an ear that initially collects sound
40. a muscle cell or gland cell that performs the body's responses to stimuli
42. the type of joint that enables mammals to rotate the forearm at the elbow
44. the ventral part of the vertebrate forebrain that helps maintain homeostasis
45. the type of system that forms a communication and coordinating network among all parts of an animal's body
46. the type of muscle filament made from actin
48. the conduction of signals from the central nervous system to effector cells
51. the structure that connects the middle ear to the pharynx
52. another name for an association neuron
53. a hard inner skeleton
55. the overall feeling of sadness and loss of interest in pleasurable activities
57. vision problem that occurs when the focal point of the lens is behind the retina
58. the abbreviation for the portion of the nervous system that includes sensory and motor neurons
59. the dorsal hollow nerve cord in vertebrates
60. the vision problem that occurs when the focal point of the lens is in front of the retina
64. a motor neuron and all the muscle fibers it controls
65. the type of arthritis that is an autoimmune disease
66. the region of the vertebrate ear that conveys vibrations from the eardrum to the oval window

ACROSS

67. the type of input from sensory receptors to integration centers in the CNS
69. an insulating layer around an axon
73. the lowest part of the vertebrate brain that controls autonomic, homeostatic functions
75. one of the three main regions of the ear where sound is first collected
76. the type of system consisting of all the muscles in the body
77. the type of neurons that receive information from sensory receptors
79. a photoreceptor in the vertebrate retina that is sensitive to black and white
80. the division of the motor nervous system composed of motor neurons that carry signals to skeletal muscles
81. the division of the autonomic nervous system that generally enhances body activities that gain and conserve energy
83. the type of potential that varies with the strength of the stimulus
84. the conversion of a stimulus signal into an electrical signal by a sensory receptor cell
86. a tough, white outer layer of connective tissue that forms the outside of the vertebrate eye
87. the type of potential that results from the voltage across the plasma membrane of a resting neuron
88. the division of the autonomic nervous system that increases energy expenditure and prepares the body for action
90. the type of body system that provides support and protection for the body
91. a bone disorder marked by thinner and more easily broken bones
92. a relay point between two neurons or between a neuron and an effector cell

DOWN

1. the main eye structure that focuses light onto the retina
2. the largest and most dominant portion of the vertebrate forebrain
4. the abbreviation for the brain and spinal cord in vertebrate animals
5. the type of muscle filament made from myosin
6. a rapid change in the membrane potential of an excitable cell
7. strong, fibrous tissues that join bones to muscles
10. the portion of a neuron that carries nerve impulses away from the cell body
11. the hindbrain and midbrain of the vertebrate central nervous system
16. a narrow gap separating the synaptic knob of a transmitting neuron from a receiving neuron or an effector cell
17. the surface of the cerebrum
20. the part of the skeleton associated with the skull and vertebrae
21. the part of a neuron that houses the nucleus
24. the part of nervous system that consists of the sympathetic and parasympathetic divisions
26. a contractile thread in a muscle cell made of many sarcomeres
28. a photoreceptor in the vertebrate eye that detects color
29. a thick band of nerve fibers that interconnects the cerebral hemispheres
30. a short, highly branched process of a neuron that conveys nerve impulses toward the cell body
33. a sheet of tissue that separates the outer ear from the middle ear
34. the type of receptors that respond to electromagnetic energy
35. the master control center of the central nervous system
37. a broad band of nerve fibers that connects the sides of the cerebellum and medulla oblongata

DOWN

41. the type of cells in the nervous system that protect, insulate, and reinforce a neuron
43. the jellylike material that fills the posterior cavity of the vertebrate eye
47. the type of joint that permits movement in a single plane
49. the minimum change in a membrane's voltage that must occur to trigger an action potential
50. strong, fibrous tissues that join bones
54. a sensory receptor that detects physical deformations in the body's environment
56. a phenomenon that occurs when the two hemispheres of the brain become specialized for different functions
61. the innermost layer of the vertebrate eye containing rods and cones
62. a sophisticated integrating center in the forebrain located just above the hypothalamus
63. the type of receptors that detect injury
68. a sensory receptor that detects heat or cold
70. another name for a nerve cell
71. a chemical messenger released from the synaptic terminal of a neuron at a chemical synapse
72. the type of neuron that conveys signals from the CNS to effector cells
74. a receptor of light
78. the fundamental unit of muscle contraction
82. any factor that causes a nerve signal to be generated
85. the type of muscle generally responsible for voluntary movements
89. the opening in the iris

CHAPTER 28

The Life of a Flowering Plant

Studying Advice

a. Have you ever considered how much plants are a part of our daily lives? Look around you right now. How many plant products do you see? This book page is made of paper, derived from wood. Perhaps you have a pencil nearby with a wooden shaft? You might have a bag of chips or other plant products around for a snack (nuts or raisins). The curtains or material covering the furniture nearby are likely covered with plant-derived materials (such as cotton), including the clothing you are currently wearing. Plants and their products are all around us. This chapter is about their parts, what they do, and what we do with them. What would we do without them?

b. This chapter should be read before Chapter 29. The figures in this chapter and the following organizing tables are especially helpful in organizing, understanding, and learning this information.

Student Media

Activities

Root, Stem, and Leaf Sections
Primary and Secondary Growth
Angiosperm Life Cycle
Seed and Fruit Development

BioFlix

Tour of a Plant Cell

BLAST Animations

Flower Structure
Pollination and Fertilization

MP3 Tutors

From Flower to Fruit

Process of Science

How Are Trees Identified By Their Leaves?

What Are Functions of Monocot Tissues?

What Tells Desert Seeds When to Germinate?

Videos

Discovery Channel Video: Colored Cotton

Discovery Channel Video: Plant Pollination

Root Growth in a Radish Seed (time-lapse)

Flower Blooming (time-lapse)

Flowering Plant Life Cycle (time-lapse)

Bat Pollinating Agave Plant

Bee Pollinating

Organizing Tables

TABLE 28.1 Compare the structure of monocots and dicots in the table.

	Monocots	Dicots
The number of seed leaves		
Pattern of veins in the leaves		
The number of flower petals and other parts		
Root systems		
Examples		

TABLE 28.2 Compare the structures and functions of the following types of plant cells.

	Structure	Function
Parenchyma cells		
Collenchyma cells		
Sclerenchyma cells		
Water-conducting cells		
Food-conducting cells		

TABLE 28.3 Compare annuals, biennials, and perennials in the table.

	Typical Number of Years That They Live	Examples
Annuals		
Biennials		
Perennials		

Content Quiz

Directions: Identify the *one* best answer for the multiple-choice questions. For true/false questions, determine if the statement is true or false. If false, change the underlined word(s) to make the statement true. Finally, add the correct word(s) to the fill-in-the-blank questions to make the statements true.

Biology and Society: Plants and Human Civilization

1. Which one of the following statements is *false*? Wheat:
   A. accounts for about 20% of all calories consumed worldwide.
   B. was cultivated in Egypt, India, China, and northern Europe about 7,000 years ago.
   C. yield was doubled during the green revolution.
   D. is one of the most valuable cash crops in the United States.
   E. is used for fuel and clothing in many parts of the world.

2. True or False? Flowering plants make up over 90% of the plant kingdom.

3. True or False? The domestication of wild animals allowed for the surplus production of food and the formation of year-round farming villages.

4. The _____________ revolution doubled the wheat yield while reducing the cost by half.

The Structure and Function of a Flowering Plant

MONOCOTS AND DICOTS

5. Which one of the following statements is *false*?
   A. Most monocots have leaves with parallel veins.
   B. Monocots have flowers with petals and other parts in multiples of three.
   C. Monocots have a large, vertical root.
   D. Monocots have stems with vascular tissues arranged in a complex array of bundles.
   E. Monocots include the orchids, palms, lilies, grains, and other grasses.

6. Which one of the following statements is *false*?
   A. Dicots have leaves with a multibranched network of veins.
   B. Dicots typically have a taproot.
   C. Dicots have flowers with petals and other parts in multiples of four or five.
   D. Dicot stems have vascular bundles arranged in a scattered pattern.
   E. Dicots include shrubs, trees, and many of our food crops.

7. True or False? Dicot embryos typically have <u>one leaf</u>.

8. The first leaves that appear on a plant embryo are _____________.

PLANT ORGANS: ROOTS, STEMS, AND LEAVES

9. Which one of the following is *not* an evolutionary adaptation that made it possible for plants to move onto land? The ability to:
   A. use photosynthesis.
   B. absorb light and take in carbon dioxide from the air.
   C. take up water and minerals from the soil.
   D. survive dry conditions.

10. Which one of the following is *not* a function of a plant's root system?
    A. anchors the plant in the soil
    B. produces sugar
    C. transports minerals and water
    D. absorbs minerals and water
    E. stores food

11. True or False? In a tree, the trunk and branches are its <u>roots</u>.

12. True or False? Grasses and most other monocots have long leaves <u>without</u> petioles.

13. True or False? The spines of a barrel cactus are modified <u>stems</u>.

14. Carrots and turnips are examples of _____________, which store food.

15. Stems, leaves, and adaptations for reproduction are all part of the _____________ system of a plant.

16. In a plant stem, the points at which leaves are attached are called _____________, and the portions of the stem in between are called _____________.

17. In the process called _____________, the terminal bud of a plant produces hormones that inhibit growth of its axillary buds.

18. Removing the terminal buds of some plants stimulates the growth of _____________ buds.

19. Using asexual reproduction, a strawberry plant produces a new plant at the tip of its _____________.

20. A sweet potato is a modified _____________ but a white potato is a modified _____________.

21. Stems and leaves depend on the water and minerals absorbed by _____________.

22. Located near root tips, tiny projections called _____________ increase the surface area for absorption of water and minerals.

23. The primary sites of photosynthesis in most plants are the _____________.

24. A sweet pea plant uses a modified leaf called a(n) _____________ to climb up its supports.

PLANT TISSUES AND TISSUE SYSTEMS

Matching: Match each item on the left to its best description on the right.

_____ 25. phloem

_____ 26. xylem

_____ 27. epidermis

_____ 28. cuticle

_____ 29. endodermis

_____ 30. pith

_____ 31. guard cells

_____ 32. mesophyll

_____ 33. cortex

_____ 34. stomata

A. the first defense against physical damage and infections

B. a waxy coating

C. site of food storage and water and mineral absorption in a root

D. regulate the size of the stomata

E. cells that conduct water and dissolved minerals

F. fills the center of the stem in dicots

G. tiny pores between two specialized epidermal cells

H. the ground tissue system of a leaf

I. food-conducting cells

J. the innermost layer of cortex

35. True or False? <u>All</u> plant stems have vascular tissue systems arranged in numerous vascular bundles.

36. True or False? Each vein in a leaf is a vascular bundle composed of xylem and phloem surrounded by <u>parenchyma</u> cells.

37. Made up of xylem and phloem, the _____________ tissue system provides support and transports water and nutrients throughout the plant.

38. The _____________ tissue system has diverse functions, including photosynthesis, storage, and support.

39. The _____________ tissue system of plants covers and protects leaves, stems, and roots.

PLANT CELLS

Matching: Match each item on the left to its best description on the right.

_____ 40. chloroplasts

_____ 41. central vacuoles

_____ 42. cell walls

_____ 43. parenchyma cells

_____ 44. collenchyma cells

_____ 45. food-conducting cells

_____ 46. water-conducting cells

_____ 47. sclerenchyma cells

A. provide support in growing parts of plant

B. cell part made mainly of cellulose

C. form chains with overlapping ends, are dead when mature

D. help maintain turgor in a cell

E. form chains with overlapping ends, are alive when mature

F. organelles that contain photosynthetic pigments

G. their rigid cell walls support the plant as steel beams support a building

H. most abundant type of cell in most plants

48. True or False? A distinctive feature of plant cells is their vacuole surrounding the plasma membrane.

49. Plant cell walls are made mainly of the structural carbohydrate _____________.

50. Long water-conducting cells with tapered ends are called _____________; wider, shorter, and less tapered cells are called _____________.

Plant Growth

51. Which of the following is a characteristic of most plants but is rarely seen in animals?
    A. sexual reproduction
    B. asexual reproduction
    C. the formation of tissues
    D. indeterminate growth
    E. mitosis

52. True or False? Plants that typically live only for two years produce flowers in the first year.

53. Plants that live and reproduce for many years are called _____________.

54. Corn, wheat, and rice typically live for just one year and are called _____________.

PRIMARY GROWTH: LENGTHENING

55. Which one of the following statements about meristems is *false*?
    A. Meristems consist of unspecialized cells that divide and generate new cells and tissues.
    B. Meristems are present only during the embryonic stages of a plant's life.
    C. Cell division in the apical meristems of roots and shoots contributes to primary growth.
    D. Growth in all plants is made possible by meristems.

56. True or False? Cells produced by secondary growth form tissues that develop into the epidermis, cortex, and vascular tissue.

57. At the tip of a root is a(n) _____________, a thimble-like cone of cells that protects the apical meristem.

58. Meristems at the tips of roots and in the terminal and axillary buds of shoots are called _____________.

59. Cell division in the apical meristems of roots and shoots produces _____________ growth.

SECONDARY GROWTH: THICKENING

60. Secondary growth involves cell division in two meristems:
    A. the vascular cambium and the cork cambium.
    B. the vascular cambium and the apical meristems.
    C. the cork cambium and the apical meristems.
    D. wood rays and apical meristems.
    E. the cork cambium and rays.

61. Which one of the following statements about growth rings is *false*?
    A. Annual growth rings result from layers of uneven secondary xylem growth due to seasonal variations.
    B. Early wood cells are usually smaller in diameter and thicker-walled than those produced in summer.
    C. Each tree ring consists of a cylinder of early wood surrounded by a cylinder of summer wood.
    D. The vascular cambium becomes dormant each year during winter.

62. After several decades of growth by a tree, which one of the following is typically dead?
    A. the vascular cambium
    B. the cork cambium
    C. the youngest secondary phloem
    D. mature cork cells

63. True or False? Stems and roots often thicken through primary growth.

64. True or False? After several decades of secondary growth, the bulk of a tree trunk is dead tissue.

65. During secondary growth, the _____________ first appears as a cylinder of actively dividing cells between the primary xylem and primary phloem.

66. Yearly production of a new layer of secondary _____________ accounts for most of the growth in thickness of a perennial plant.

67. Based upon an analysis of tree rings, it now appears that the lost colony of Roanoke suffered from a severe _____________.

68. Cork is produced by a meristem called _____________.

69. Everything external to the vascular cambium (the secondary phloem, cork cambium, and cork) is called _____________.

The Life Cycle of a Flowering Plant

THE FLOWER

Matching: Match each item on the left to its best description on the right.

_____ 70. petals

_____ 71. stamens

_____ 72. ovule

_____ 73. sepals

_____ 74. carpel

_____ 75. ovary

_____ 76. style

_____ 77. anther

_____ 78. stigma

A. has a longer slender neck and stigma at its tip

B. the receiving surface for pollen grains

C. a stalk tipped by an anther

D. it leads to the ovary at the base of the stigma

E. often bright and colorful, they advertise a flower to pollinators

F. contains the developing egg and cells that support it

G. a terminal sac on the stamen where pollen develops

H. enclose and protect the flower bud

I. houses reproductive structures called the ovules

79. True or False? A flower's main parts, the sepals, petals, stamens, and carpels, are modified stems.

80. True or False? Many plants can reproduce sexually and asexually.

81. True or False? Using sexual reproduction, a single plant can produce many identical plants quickly and efficiently.

82. In angiosperms, the organ specific to sexual reproduction is the _____________.

83. The term _____________ is sometimes used to refer to a single carpel or a group of fused carpels.

OVERVIEW OF THE FLOWERING PLANT LIFE CYCLE, POLLINATION, AND FERTILIZATION

Matching: Match each item on the left to its best description on the right.

_____ 84. sporophyte

_____ 85. gametophyte

_____ 86. pollen grain

_____ 87. embryo sac

_____ 88. pollination

_____ 89. fertilization

A. a plant's haploid generation

B. the female gametophyte

C. the male gametophyte

D. the diploid plant body

E. the delivery of pollen to the stigma of a carpel

F. the union of male and female gametes

90. True or False? The pollen grain germinates on the <u>stamen</u>.

91. The formation of a zygote and a cell with a triploid nucleus is called _____________.

92. The triploid cell formed during fertilization will give rise to a food-storing tissue called _____________.

SEED FORMATION

93. After a mature seed is produced:
    A. the endosperm disintegrates.
    B. the seed coat splits open.
    C. the zygote divides by meiosis.
    D. the embryo stops developing.
    E. an embryonic root emerges.

94. True or False? The <u>cotyledon</u> encloses and protects the endosperm.

95. The _____________ is a multicellular mass that nourishes the embryo until it becomes a self-supporting seedling.

FRUIT FORMATION, SEED GERMINATION

96. Which one of the following statements about fruits is *false*?
    A. Fruits are highly variable.
    B. A corn kernel is a fruit.
    C. A pea pod is a fruit.
    D. Fruits develop after the seeds develop.
    E. A fruit houses and protects seeds and helps disperse them from the parent plant.

97. Germination is typically triggered by:
    A. mechanical damage to the seed coat.
    B. fertilization.
    C. destruction of the endosperm.
    D. ingestion by an animal.
    E. the absorption of water by the seed.

98. True or False? The first thing to emerge from a germinating seed is the embryonic shoot.

99. True or False? In the wild, only a few seedlings endure long enough to reproduce.

100. Fleshy, edible fruits entice animals that help spread ___________.

Evolution Connection: The Problem of the Disappearing Bees

101. Most angiosperms depend on insects, birds, or mammals for:
    A. germination.
    B. fertilization.
    C. pollination and seed dispersal.
    D. nutritional support of the developing plant embryo.
    E. elimination of their competition.

102. True or False? Colony collapse disorder is a poorly understood threat to bee colonies.

103. Flowers pollinated by bees often have markings that reflect ___________ light.

Word Roots

a = without (asexual reproduction: reproduction without sex)
apic = the tip; **meristo** = divided (apical meristems: embryonic plant tissue on the tips of roots and in the buds of shoots that supplies cells for the plant to grow)
bienn = every two years (biennial: a plant that requires two years to complete its life cycle)
chloro = green (chloroplasts: the green, photosynthetic organelle common in plants)
coll = glue; **enchyma** = an infusion (collenchyma cells: a flexible plant cell type that occurs in strands or cylinders that support young parts of the plant without restraining growth)
di = two (dicot: a flowering plant whose embryos have two seed leaves)
endo = inner; **derm** = skin (endodermis: the innermost layer of the cortex in plant roots)
epi = over (epidermis: the dermal tissue system in plants; the outer covering of animals)
gamet = a wife or husband (gametophyte: the multicellular haploid form in organisms undergoing alternation of generations, which mitotically produces haploid gametes that unite and grow into the sporophyte generation)

inter = between (internode: the segment of a plant stem between the points where leaves are attached)

meso = middle; **phyll** = a leaf (mesophyll: the ground tissue of a leaf, sandwiched between the upper and lower epidermis and specialized for photosynthesis)

mono = one (monocot: a flowering plant whose embryos have a single seed leaf)

perenni = through the year (perennial: a plant that lives for many years)

phloe = the bark of a tree (phloem: the portion of the vascular system in plants consisting of living cells arranged into elongated tubes that transport sugar and other organic nutrients throughout the plant)

sclero = hard (sclerenchyma cell: a supportive cell with a very rigid secondary wall)

sporo = a seed; **phyto** = a plant (sporophyte: the multicellular diploid form in organisms undergoing alternation of generations that results from a union of gametes and that meiotically produces haploid spores that grow into the gametophyte generation)

stam = standing upright (stamen: the pollen-producing male reproductive organ of a flower, consisting of an anther and filament)

stoma = mouth (stomata: a hole in plant leaves through which water, oxygen, and carbon dioxide pass)

vascula = little tube (vascular tissue system: a transport system of tubes for water and nutrients in plants)

xyl = wood (xylem: the tube-shaped, nonliving portion of the vascular system in plants that carries water and minerals from the roots to the rest of the plant)

Key Terms

angiosperms
annuals
anther
apical dominance
apical meristem
asexual reproduction
axillary buds
bark
biennials
blade
carpel
cell wall
central vacuole
chloroplasts
collenchyma cells
cork
cork cambium
cortex
cotyledons
cuticle
dermal tissue system
determinate growth
dicot
double fertilization
embryo sac
endodermis
endosperm
eudicots
fertilization
flower
fruit
gametophyte
germinates
ground tissue system
guard cells
indeterminate growth
internodes
leaves
lignin
meristem
mesophyll
monocot
nodes
organ
ovary
ovule
parenchyma cells
perennials
petals
petiole
phloem
pistil
pith
pollen grain
pollination
primary growth
root
root cap
root hairs
root system
sclerenchyma cells
secondary growth
seed
seed coat
sepals
sexual reproduction
shoot system
sporophyte
stamen
stems
stigma
stomata
terminal bud
tissue
tissue system
tracheids
tubers
vascular cambium
vascular tissue system
vessel elements
wood
xylem

Crossword Puzzle

Use the Key Terms list from this chapter to fill in the crossword puzzle.

ACROSS

1. a tiny projection growing just behind a root tip; it increases the surface area of a root
3. the diploid plant body
4. secondary xylem of a plant
8. the starch-storing portion of a potato plant
9. when mature, these types of cells are dead, and their rigid cell walls support the plant much like steel beams in a building
10. a plant that lives for many years
11. the pollen-producing male reproductive organ of a flower
13. the aerial portion of a plant body, consisting of stems, leaves, and flowers
16. plant tissues are organized into these
17. vascular tissue that contains food-conducting cells
19. a structure consisting of several tissues adapted as a group to perform specific functions
20. the type of reproduction that results in the creation of genetically distinct offspring by the fusion of two haploid gametes
21. the ground tissue of a leaf
23. the type of growth characteristic of plants, in which the organism grows as long as it lives
26. the type of specialized epidermal plant cells that regulate the size of stomata
27. a plant structure that anchors the plant in the soil, absorbs and transports minerals and water, and stores food
28. the tube-shaped, nonliving portion of the vascular system in plants that carries water and minerals
31. the placement of pollen onto the stigma of a carpel by wind or animal carriers
32. the outermost protective layer of a plant's bark
33. a point along the stem of a plant where a leaf is attached
34. all of a plant's roots that anchor it in the soil, absorb and transport minerals and water, and store food
36. the ground tissue system of a root, which stores food and absorbs minerals
38. the main site of photosynthesis in most plants
41. all of the plant tissues external to the vascular cambium
43. in an angiosperm, a short stem with four sets of modified leaves, bearing structures that function in sexual reproduction
44. the type of tissue system consisting mostly of parenchyma cells that makes up the bulk of a young plant
45. an embryonic shoot present in the angle formed by a leaf and stem
46. the type of dominance in which the terminal bud produces hormones that inhibit growth of axillary buds
47. in flowers, the portion of a carpel in which the egg-containing ovules develop
48. a part of the ground tissue system that fills the center of a stem and is often important in food storage
49. a modified leaf of a flowering plant. Petals are the often colorful parts of a flower that advertise it to insects and other pollinators
53. the general name for flowering plants
54. the type of fertilization in angiosperms in which two sperm cells unite with two cells in the embryo sac to form the zygote and endosperm
57. the female reproductive organ of a flower
59. seed leaves
60. a plant that completes its entire life cycle in a single year or growing season
62. the type of tissue system that covers and protects leaves, stems, and roots
64. a waxy coating on the surface of stems and leaves that helps retain water
65. the sticky part of a flower's carpel
66. members of a group consisting of the vast majority of flowering plants that have two cotyledons

67. a nutrient-rich tissue that provides nourishment to the developing embryo in angiosperm seeds
68. what a seed does when it first begins to grow
70. a large, fluid-containing organelle in plant cells
72. a tough protective layer that encloses the endosperm
73. the type of growth that ends after an organism reaches a certain size
74. a continuous cylinder of meristematic cells surrounding the xylem and pith that produces secondary xylem and phloem

DOWN

1. a cone of cells at the tip of a plant root that protects the apical meristem
2. type of growth characterized by an increase in girth of the stems and roots
5. whorls of modified leaves that enclose and protect the flower bud before it opens
6. the most abundant type of cell in most plants
7. the stalk of a leaf, which joins the leaf to the stem
8. long and narrow, porous, water-conducting cells
12. wider and shorter, open-ended, water-conducting cells
14. an integrated group of similar cells that perform a specific function within a multicellular organism
15. the segment of a plant stem between the points where leaves are attached
17. in a seed plant, the male gametophytes that develop within the anthers of stamens
18. a subdivision of flowering plants whose members possess one embryonic seed leaf
22. a chemical component of wood
24. the multicellular haploid form in organisms undergoing alternation of generations
25. the type of growth initiated by the apical meristems of a plant root or shoot
29. the type of flexible plant cell type that occurs in strands or cylinders that support young parts of the plant without restraining growth
30. an embryonic tissue at the tip of a shoot
35. a structure that develops in the plant ovary and contains the female gametophyte
37. unspecialized cells that divide and generate new cells and tissues
39. the innermost layer of the cortex in plant roots
40. the flattened portion of a leaf
42. the type of meristem found at the tips of roots and in the terminal and axillary buds of shoots
46. the terminal pollen sac of a stamen
48. a modified leaf of a flowering plant that is often the most colorful part of a flower
50. a type of system formed by xylem and phloem throughout a plant, serving as a transport system for water and nutrients
51. the union of haploid gametes to produce a diploid zygote
52. the type of reproduction that creates offspring derived from a single parent
54. a subdivision of flowering plants whose members possess two embryonic seed leaves
55. the female gametophyte contained in the ovule of a flowering plant
56. a plant that requires two years to complete its life cycle
58. a microscopic pore surrounded by guard cells in the epidermis of leaves and stems
59. meristematic tissue that produces cork cells during secondary growth of a plant
61. sites of photosynthesis in plant cells
63. a mature ovary of a flower that protects dormant seeds and aids in their dispersal
69. a plant embryo packaged with a food supply within a protective covering
71. a part of a plant's shoot system that supports the leaves and reproductive structures

1
2
3
4
5
6
7
8
9
10
11
12
13
14
15
16
17
18
19
20
21
22
23
24
25
26
27
28
29
30
31
32
33
34
35
36
37
38
39
40
41
42
43
44
45
46
47
48
49
50
51
52
53
54
55
56
57
58
59
60
61
62
63
64
65
66
67
68
69
70
71
72
73
74

CHAPTER 29

The Working Plant

Studying Advice

a. Have you ever wondered how they produce seedless grapes? Have you ever struggled to keep a plant alive, wondering what you were doing wrong? Have you ever wondered how plants grow toward sunlight? This chapter reveals why and how plants do what they do and makes caring for plants a little easier.

b. Chapter 29 builds upon the content in Chapter 28. If you have not already read Chapter 28, be prepared to read at least sections of it as background for Chapter 29.

Student Media

Activities

How Plants Obtain Minerals from Soil

Transport of Xylem Sap

Translocation of Phloem Sap

Leaf Abscission

Flowering Lab

BioFlix

Water Transport in Plants

Graph It

Global Soil Degradation

LabBench

Transpiration

MP3 Tutors

Transpiration

Process of Science

How Does Acid Precipitation Affect Mineral Deficiency?
How Is the Rate of Transpiration Calculated?
What Plant Hormones Affect Organ Formation?

Videos

Discovery Channel Video: Trees
Sun Dew Trapping Prey
Root Growth in a Radish Seed
Phototropism
Gravitropism
Mimosa Leaf

Organizing Tables

TABLE 29.1 Distinguish between micronutrients and macronutrients in the table.

	Definition	Examples
Micronutrients		
Macronutrients		

TABLE 29.2 Identify the main functions of the five plant hormones in the table.

Plant Hormones	Functions
Auxin	
Ethylene	
Cytokinins	
Gibberellins	
Abscisic Acid	

TABLE 29.3 Distinguish between the three types of tropisms in the table.

	Definition	Examples
Phototropism		
Thigmotropism		
Gravitropism		

Content Quiz

Directions: Identify the *one* best answer for the multiple-choice questions. For true/false questions, determine if the statement is true or false. If false, change the underlined word(s) to make the statement true. Finally, add the correct word(s) to the fill-in-the-blank questions to make the statements true.

Biology and Society: Planting Hope in the Wake of Katrina

1. Which of the following provides a cost-effective and efficient way to remove toxins from soil?
   A. digging up contaminated topsoil and hauling it away
   B. spraying contaminated topsoil with diluted bleach solutions
   C. using contaminated topsoil to raise vegetable gardens for people to eat
   D. raising sunflowers on contaminated topsoil and hauling mature sunflowers away
   E. waiting until several years of rains have washed away the toxins in the contaminated topsoil

2. True or False? Plants have evolved amazing abilities to pull <u>water and nutrients</u> out of the soil and air.

3. The biggest obstacle to the sunflower approach to removing toxins from soil is ______________.

How Plants Acquire and Transport Nutrients

PLANT NUTRITION

4. Which one of the following is a micronutrient?
   A. carbon
   B. oxygen
   C. hydrogen
   D. nitrogen
   E. iron

5. Which one of the following is a macronutrient?
   A. manganese
   B. copper
   C. sulfur
   D. chlorine
   E. zinc

6. A plant needs only minute quantities of micronutrients because it:
   A. can easily get them from the soil.
   B. recycles micronutrients.
   C. can produce them from macronutrients.
   D. only needs them a few days of the year.
   E. needs them only to reproduce.

7. True or False? Plants require relatively large amounts of <u>micronutrients</u>.

8. True or False? A deficiency of any micronutrient <u>can kill</u> a plant.

9. One of the fastest-growing segments of agriculture in the United States, ____________ farming intends to build a sustainable agriculture system.

10. An element is considered a(n) ____________ plant nutrient if the plant must obtain it to complete its life cycle.

11. An organic fertilizer composed of chemically complex organic matter, ____________ is a soil-like mixture.

12. There are two main types of ____________, inorganic and organic.

13. The most common nutritional problem for plants is a shortage of ____________.

FROM THE SOIL INTO THE ROOTS

14. Which of the following helps a plant regulate the mineral composition of its vascular system?
    A. the cuticle of the leaves
    B. the plasma membranes of root cells
    C. parenchyma cells of the stems
    D. mycorrhiza
    E. sunlight striking the surfaces of leaves

15. True or False? <u>All</u> substances that enter a plant root are dissolved in water.

16. Roots efficiently extract nutrients from soil due to their ____________, extensions of epidermal cells that dramatically increase the surface available for absorption.

17. Most plants gain absorptive surface through mutually beneficial symbiotic associations with fungi in an association called ____________.

THE ROLES OF BACTERIA IN NITROGEN NUTRITION

Matching: Match each item on the left to its best description on the right.

_____ 18. nitrogen-fixing bacteria

_____ 19. ammonifying bacteria

_____ 20. nitrifying bacteria

_____ 21. legumes

_____ 22. root nodules

_____ 23. nitrogen fixation

A. add ammonium to soil by decomposing organic matter

B. bacteria that convert atmospheric N_2 to ammonium

C. plants that produce their seeds in pods

D. the process of converting atmospheric N_2 to ammonium

E. convert soil ammonium to nitrate

F. plant root cells containing vesicles filled with bacteria

24. True or False? Plants <u>can</u> use the form of nitrogen found in air.

25. For plants to absorb ____________ from the soil, it must first be converted to ammonium ions (NH_4^+) or nitrate ions (NO_3^-).

THE TRANSPORT OF WATER

26. Transpiration exerts a pull on a tense string of water molecules that are held together by ____________ and helped upward by ____________.
    A. cohesion; adhesion
    B. adhesion; cohesion
    C. cohesion; gravity
    D. adhesion; gravity
    E. gravity; cohesion

27. Which one, if any, of the following environmental conditions would *not* increase the amount of transpiration in a plant?
    A. intense sunlight
    B. warm weather
    C. high humidity
    D. windy conditions
    E. All of the above would increase the amount of transpiration in a plant.

28. Which one of the following statements is *false*?
    A. Stomata are usually open during the day.
    B. Stomata are usually open at night to allow oxygen to enter from the atmosphere.
    C. During the day, CO_2 can enter the leaf from the atmosphere.
    D. Stomata may close during the day if a plant is losing water too fast.
    E. Two guard cells flanking each stoma control its opening by changing shape.

29. True or False? The sticking together of molecules of the same kind defines <u>adhesion</u>.

30. True or False? The transport of xylem sap requires <u>no energy</u> expenditure by the plant.

31. Mature water-conducting cells of ______________ conduct xylem sap from a plant's roots to the tips of its leaves.

THE TRANSPORT OF SUGARS

32. Which one of the following statements about the pressure-flow mechanism is *false*?
    A. Each food-conducting tube in phloem tissue has a source end and a sink end.
    B. At source, sugar is loaded from a photosynthetic cell into a phloem tube.
    C. At the sugar sink, water and sugar enter the phloem tube.
    D. At the sugar sink, the exit of water lowers the water pressure in the tube.
    E. Phloem sap always flows from a sugar source to a sugar sink.

33. True or False? A sugar <u>sink</u> is a location in a plant where sugar is being produced.

34. True or False? <u>Phloem sap</u> moves freely from the cytoplasm of one cell to the next.

35. The main function of ______________ is to transport sugars produced by a plant using photosynthesis.

Plant Hormones

AUXINS, ETHYLENE, CYTOKININS, GIBBERELLINS, ABSCISIC ACID

Matching: Match each hormone on the left to its best description on the right.

_____ 36. auxin	A. promotes seed germination
_____ 37. ethylene	B. promotes cell elongation in stems
_____ 38. cytokinin	C. promotes fruit ripening and dropping of leaves
_____ 39. gibberellin	D. inhibits growth
_____ 40. abscisic acid	E. promotes cytokinesis

41. Which one of the following statements about plant hormones is *false*?
   A. Plants produce hormones in very large amounts.
   B. Each type of hormone can produce a variety of effects.
   C. Small amounts of hormones can have profound effects on target cells.
   D. Hormones play critical roles in the development of plants.
   E. Hormones play critical roles in the growth of plants.

42. Which one of the following statements about auxin is *false*?
   A. Auxin promotes cell elongation in stems only within a certain concentration range.
   B. Auxin is produced by developing seeds and promotes the growth of fruit.
   C. The same concentration of auxin that promotes cell elongation in stems inhibits cell elongation in roots.
   D. Auxin is the hormone most associated with the loss of leaves in autumn.
   E. Auxin promotes growth in stem diameter.

43. Which one of the following hormones enters the shoot system from the roots and counters the inhibitory effects of auxin coming down from the terminal buds?
   A. gibberellins
   B. auxin
   C. ethylene
   D. cytokinins
   E. None of the above.

44. Which one of the following pairs stimulates cell elongation and cell division in stems and influences fruit development?
   A. gibberellins and auxin
   B. cytokinins and ethylene
   C. abscisic acid and ethylene
   D. auxin and abscisic acid
   E. cytokinins and abscisic acid

45. True or False? An uneven distribution of auxin causes the dark side of a shoot to grow slower than the light side.

46. True or False? The same hormone may have different effects at different concentrations in the same target cell.

47. True or False? Animals and plants use hormones to regulate their internal activities.

48. True or False? Some growers slow the ripening of apples by storing apples in containers flushed with ethylene.

49. A chemical signal produced in one part of the body and transported to other parts, where it acts on target cells to change their functioning, is a(n) ______________.

50. Spraying unfertilized plants with synthetic gibberellin and ______________ can make apples, currants, and eggplants develop without pollination and seed production.

51. The growth of a shoot toward light is called ______________.

Response to Stimuli

TROPISMS, PHOTOPERIOD

52. Long-night plants respond to:
    A. warmer daytime temperatures.
    B. stronger intensities of light.
    C. the relative length of time without sunlight.
    D. phases of the moon.
    E. the total amount of sunlight through the day.

53. True or False? Short-night plants usually flower in late <u>fall or early winter</u>.

54. Growth of a plant in response to touch is called ______________.

55. Growth of a plant in response to gravity is called ______________.

56. The relative length of days and nights is called ______________.

Evolution Connection: The Interdependence of Organisms

57. Most flowering plants depend on:
    A. insects or other animals for pollination and seed dispersal.
    B. nitrogen-fixing bacteria.
    C. the fungi of mycorrhizae.
    D. the sun for photosynthesis.
    E. All of the above.

58. True or False? The fossil record shows that <u>mycorrhizae</u> have existed since plants first evolved.

59. As a tadpole, a frog feeds on ______________.

60. As an adult, a frog feeds on ______________.

Word Roots

aux = grow, enlarge (auxins: a class of plant hormones that promote cell elongation, stimulate secondary growth, and promote the development of leaf traces and fruit)

co = together; **hesita** = stick fast (cohesion: the sticking together of molecules of the same kind)

cyto = cell; **kine** = moving (cytokinins: a class of related plant hormones that retard aging and act in concert with auxins to stimulate cell division, influence the pathway of differentiation, and control apical dominance)

gibb = humped (gibberellins: a class of related plant hormones that stimulate growth in the stem and leaves, trigger the germination of seeds and breaking of bud dormancy, and stimulate fruit development with auxin)

gravi = heavy; **trop** = turn, change (gravitropism: directional growth of plants in response to gravity)

macro = large (macronutrients: elements required by plants and animals in relatively large amounts)

micro = small (micronutrients: elements required by plants and animals in very small amounts)

myco = a fungus (mycorrhizae: mutualistic associations of plant roots and fungi)

photo = light (phototropism: growth of a plant shoot toward or away from light)

thigmo = touch (thigmotropism: growth of a plant in response to touch)

Key Terms

abscisic acid
adhesion
auxin
cohesion
compost
cytokinins
essential element
ethylene
fertilizer
gibberellins
gravitropism
hormones
macronutrients
micronutrients
minerals
mycorrhiza
nitrogen fixation
organic farming
phloem sap
photoperiod
phototropism
pressure-flow mechanism
sugar sink
sugar source
thigmotropism
transpiration
transpiration-cohesion-tension mechanism
tropisms
xylem sap

Crossword Puzzle

Use the Key Terms list from this chapter to fill in the crossword puzzle.

ACROSS

6. a mutualistic association of a plant root and fungus
7. the type of nutrients that plants require in extremely small amounts
8. regulatory chemicals that travel away from their site of origin to affect internal activities
9. a type of farming that is conservation-minded and environmentally safe as well as profitable
10. the plant hormone that inhibits the germination of seeds and promotes dormancy
13. a type of element that a plant must obtain to complete its life cycle
14. a soil-like mixture of decomposed organic matter used to fertilize plants
15. the sticking together of molecules of the same kind
16. a sugary solution that moves throughout a plant in various directions
19. organic ions
20. a compound that is given to plants via soil to promote the plants' growth
21. the directional growth of a plant in relation to touch
24. a plant organ in which sugar is being produced
26. a plant organ that is a net consumer or storer of sugar
27. the relative lengths of day and night that allow plants to detect the time of year

DOWN

1. the type of mechanism used to describe the movement of xylem sap
2. growth of a plant in response to gravity
3. a family of more than 100 growth-regulating plant hormones
4. the type of nutrients that plants require in relatively large amounts
5. a growth response that makes a plant grow toward or away from a stimulus
10. any chemical substance that promotes seedling elongation
11. the sticking together of molecules of different kinds
12. growth regulators that promote cytokinesis
13. a hormone that triggers a variety of aging responses in plants, including fruit ripening and dropping of leaves
17. a solution of mostly inorganic nutrients that flows through vertical tubes from the roots to the tips of the leaves
18. the assimilation of atmospheric nitrogen by certain prokaryotes into nitrogenous compounds that can be directly used by plants
22. growth of a plant shoot toward or away from light
23. the type of mechanism by which phloem sap is transported through a plant from a sugar source to a sugar sink
25. the evaporative loss of water from a plant

1
2
3
4
5
6
7
8
9
10
11
12
13
14
15
16
17
18
19
20
21
22
23
24
25
26
27

ANSWER KEY

Chapter 1

Content Quiz Answers

1. B
2. E
3. False, genes
4. stimuli
5. A
6. False, populations
7. cells, atoms
8. A
9. C
10. True
11. biosphere
12. ecosystems
13. cell
14. genome
15. B
16. D
17. C
18. A
19. D
20. B
21. A
22. False, three
23. taxonomy
24. domains
25. E
26. D
27. B
28. True
29. Evolution
30. natural selection
31. B
32. D
33. A
34. C
35. A
36. True
37. natural selection
38. C
39. False, discovery
40. science
41. inductive
42. D
43. False, Deductive
44. hypothesis
45. A
46. discovery
47. variable
48. B
49. evolution
50. D
51. False, great/considerable
52. technology
53. C
54. B

1 BIOLOGY

2 NATURALSELECTION

3 SCIENTIFICMETHOD

4 CASESTUDY

5 HYPOTHESES

6 ECOSYSTEM

7 DISCOVERYSCIENCE

8 THEORY

9 LIFE

10 CONTROLLEDEXPERIMENT

11 HYPOTHESISDRIVENSCIENCE

Chapter 2

Content Quiz Answers

1. C
2. True
3. plaque
4. C
5. True
6. trace elements
7. D
8. C
9. D
10. True
11. protons, neutrons
12. E
13. B
14. B
15. B
16. B
17. False, oxygen atom
18. True
19. covalent
20. ions
21. D
22. True
23. products
24. A
25. B
26. A
27. A
28. A
29. True
30. solute, solution
31. water
32. A
33. False, hydrogen
34. control
35. C
36. E
37. D
38. False, 10 times
39. buffers
40. acid, base
41. D
42. True

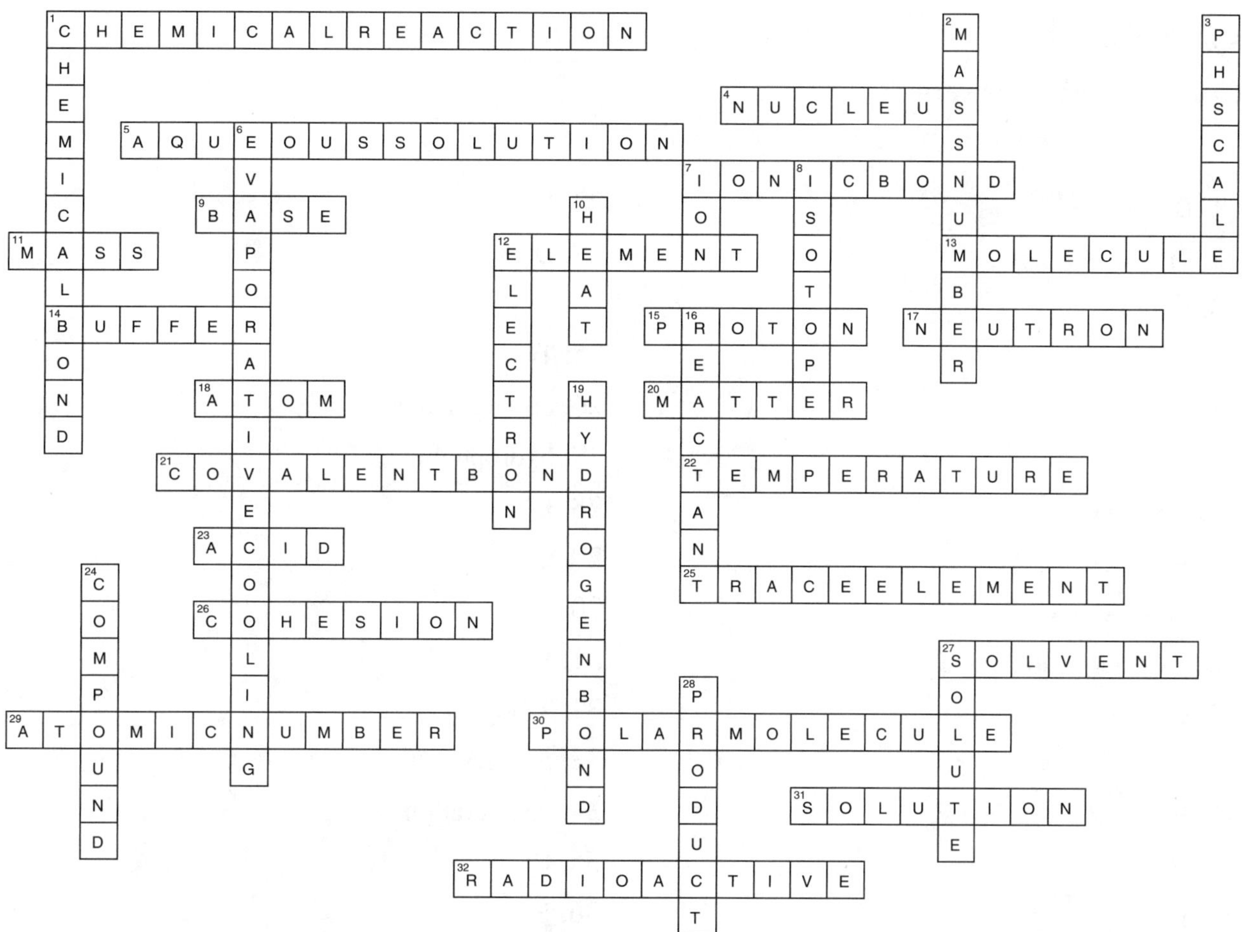

CHEMICALREACTION
MASSNUMBER
PHSCALE
NUCLEUS
AQUEOUSSOLUTION
EVAPORATIVECOOLING
IONICBOND
ISOTOPE
BASE
HEAT
MASS
ELEMENT
ELECTRON
MOLECULE
BUFFER
PROTON
REACTANT
NEUTRON
ATOM
HYDROGENBOND
MATTER
COVALENTBOND
TEMPERATURE
ACID
COMPOUND
TRACEELEMENT
COHESION
SOLVENT
SOLUTE
PRODUCT
ATOMICNUMBER
POLARMOLECULE
SOLUTION
RADIOACTIVE
CHEMICALBOND

Chapter 3

Content Quiz Answers

1. B
2. True
3. lactase
4. C
5. A
6. True
7. covalent
8. A
9. False, water
10. polymers, monomers
11. C
12. A
13. B
14. E
15. E
16. False, cellulose
17. True
18. isomers
19. monosaccharides, dehydration
20. glucose
21. E
22. C
23. D
24. E
25. True
26. atherosclerosis
27. hydrophobic
28. E
29. C
30. D
31. B
32. True
33. primary
34. denaturation
35. C
36. B
37. D
38. False, DNA
39. False, not within
40. nucleotides
41. D
42. False, least

SUGARPHOSPHATE
STEROID
MONOMER
DNA
ATHEROSCLEROSIS
HYDROLYSIS
HYDROCARBON
GENE
GLYCOGEN
CARBOHYDRATE
HYDROPHILIC
DEHYDRATIONREACTION
AMINOACID
HYDROGENATION
RNA
FUNCTIONAL
CELLULOSE
ISOMERS
MONOSACCHARIDE
FAT
DENATURATION
HYDROPHOBIC
DOUBLEHELIX
LIPID
POLYPEPTIDE
DISACCHARIDE
PROTEIN
POLYSACCHARIDE
STARCH
SATURATED
MACROMOLECULE
UNSATURATED
NUCLEOTIDE
NUCLEICACID
TRANSFATS
POLYMER
ORGANIC
PRIMARY
PEPTIDE
TRIGLYCERIDE

Chapter 4

Content Quiz Answers

1. B
2. False, mold
3. antibiotics
4. C
5. A
6. C
7. C
8. B
9. True
10. False, nucleus
11. organelles
12. 1,000
13. scanning
14. transmission
15. eukaryotic, prokaryotic
16. cell
17. pili
18. cytoplasm
19. chloroplasts
20. E
21. B
22. D
23. D
24. mosaic
25. junctions
26. D
27. A
28. True
29. messenger RNA (mRNA)
30. ribosomes
31. C
32. A
33. A
34. False, Golgi apparatus
35. lysosomes
36. rough
37. E
38. E
39. True
40. ATP
41. cristae
42. E
43. D
44. False, microtubules
45. microtubules
46. D
47. True

Chapter 5

Content Quiz Answers

1. D
2. False, enzymes
3. energy
4. B
5. True
6. conservation of energy
7. C
8. True
9. entropy
10. A
11. C
12. False, potential
13. chemical
14. B
15. True
16. calorie, Calorie
17. B
18. False, more
19. triphosphate tail
20. D
21. True
22. ADP
23. E
24. False, lower
25. metabolism
26. A
27. False, artificial
28. directed
29. D
30. False, substrates
31. substrate, active site
32. C
33. False, feedback regulation
34. inhibitors
35. A
36. E
37. True
38. energy
39. facilitated diffusion
40. B
41. C
42. C
43. False, isotonic
44. osmoregulation
45. D
46. False, active
47. active
48. D
49. False, exocytosis
50. pinocytosis
51. E
52. True
53. proteins
54. A
55. True
56. phospholipids

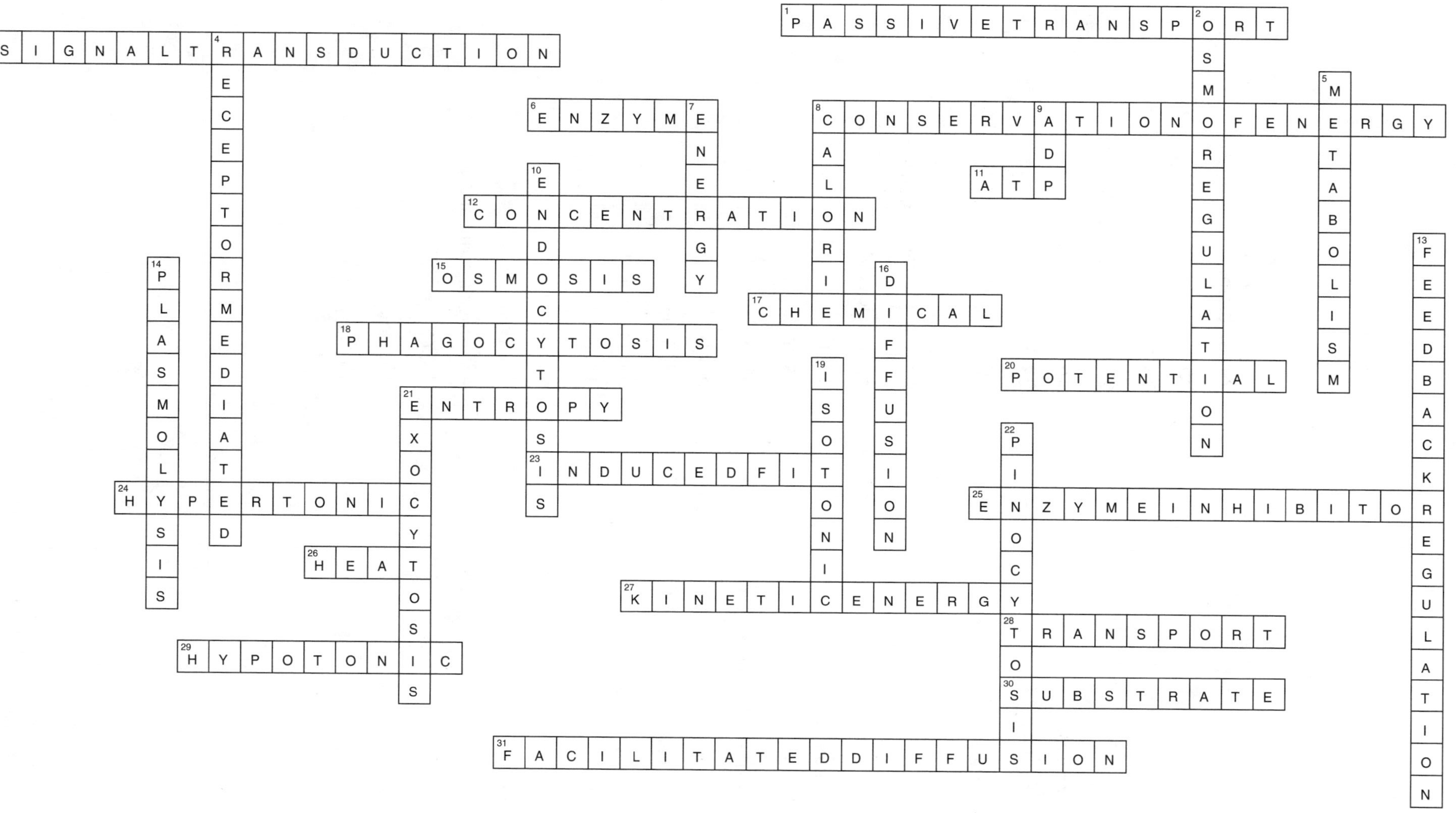

1 PASSIVETRANSPORT
2 OSMOREGULATION
3 SIGNALTRANSDUCTION
4 RECEPTORMEDIATED
5 METABOLISM
6 ENZYME
7 ENERGY
8 CONSERVATIONOFENERGY
8 CALORIE
9 ADP
10 ENDOCYTOSIS
11 ATP
12 CONCENTRATION
13 FEEDBACKREGULATION
14 PLASMOLYSIS
15 OSMOSIS
16 DIFFUSION
17 CHEMICAL
18 PHAGOCYTOSIS
19 ISOTONIC
20 POTENTIAL
21 ENTROPY
21 EXOCYTOSIS
22 PINOCYTOSIS
23 INDUCEDFIT
24 HYPERTONIC
25 ENZYMEINHIBITOR
26 HEAT
27 KINETICENERGY
28 TRANSPORT
29 HYPOTONIC
30 SUBSTRATE
31 FACILITATEDDIFFUSION

Chapter 6

Content Quiz Answers

1. B
2. True
3. fast-twitch
4. B
5. False, organic
6. False, producers
7. heterotrophs
8. A
9. D
10. D
11. False, mitochondria
12. ATP
13. water, carbon dioxide
14. D
15. False, oxygen
16. cellular respiration
17. C
18. True
19. hydrogen
20. B
21. A
22. B
23. False, oxygen
24. oxidation
25. B
26. True
27. electron transport chain
28. B
29. C
30. D
31. C
32. False, electron transport
33. acetic acid
34. E
35. True
36. 38
37. B
38. True
39. lactic acid
40. C
41. D
42. True
43. fungus, fermentation
44. C
45. True
46. oxygen

1. CITRICACIDCYCLE
2. FERMENTATION
3. HETEROTROPH
4. PRODUCER
5. PHOTOSYNTHESIS
6. NAD
7. GLYCOLYSIS
8. REDOX
9. REDUCTION
10. ATPSYNTHASE
10. AUTOTROPH
11. ANAEROBIC
12. CONSUMER
13. CELLULARRESPIRATION
14. AEROBIC
15. OXIDATION
16. ELECTRONTRANSPORT

Chapter 7

Content Quiz Answers

1. E
2. True
3. False, wood
4. photosynthesis/the sun
5. D
6. D
7. False, increases
8. stomata
9. B
10. False, water
11. glucose
12. A
13. C
14. True
15. ATP/NADPH
16. D
17. True
18. more
19. A
20. False, some
21. green
22. D
23. False, thylakoid
24. chlorophyll *a*
25. E
26. D
27. True
28. photon
29. C
30. D
31. A
32. True
33. ATP
34. A
35. True
36. stroma
37. A
38. leaves
39. stomata
40. water

1 CHLOROPLAST
2 LIGHT
3 PHOTON
4 PHOTOSYSTEM
5 NADPH
6 THYLAKOIDS
7 CHLOROPHYLLA
7 CALVIN CYCLE
8 REACTION CENTER
9 PHOTOSYNTHESIS
10 GRANA
11 STOMATA
12 CAM
13 CHLOROPHYLL
14 PRIMARY
15 ELECTROMAGNETIC
16 C3
17 C4
18 WAVELENGTH
19 STROMA

Chapter 8

Content Quiz Answers

1. E
2. True
3. cell division
4. A
5. D
6. False, mitosis
7. False, half
8. chromosomes
9. asexual
10. C
11. False, proteins
12. centromere
13. D
14. True
15. True
16. mitosis, cytokinesis
17. S
18. B
19. B
20. D
21. E
22. False, animal
23. cell plate
24. centrosomes
25. B
26. C
27. True
28. benign
29. chemotherapy
30. D
31. False, autosomes
32. homologous
33. B
34. True
35. 25
36. E
37. True
38. haploid
39. B
40. A
41. False, do
42. meiosis II
43. C
44. False, meiosis
45. True
46. chiasma
47. True
48. E
49. E
50. False, sex chromosomes
51. Down syndrome
52. C
53. True
54. sexual

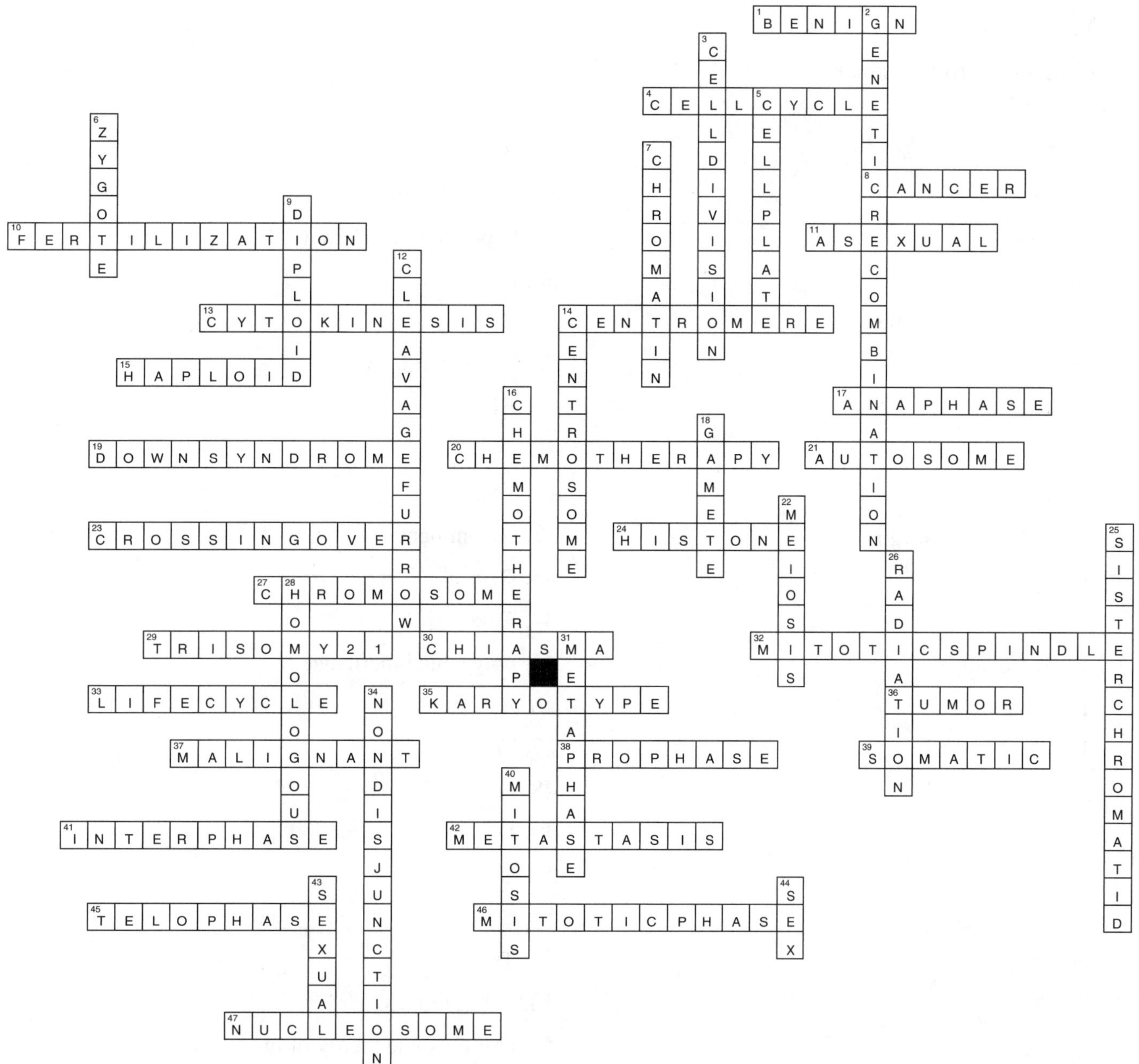
1 BENIGN
4 CELLCYCLE
8 CANCER
10 FERTILIZATION
11 ASEXUAL
13 CYTOKINESIS
14 CENTROMERE
15 HAPLOID
17 ANAPHASE
19 DOWNSYNDROME
20 CHEMOTHERAPY
21 AUTOSOME
23 CROSSINGOVER
24 HISTONE
27 CHROMOSOME
29 TRISOMY21
30 CHIASMA
32 MITOTICSPINDLE
33 LIFECYCLE
35 KARYOTYPE
36 TUMOR
37 MALIGNANT
38 PROPHASE
39 SOMATIC
41 INTERPHASE
42 METASTASIS
45 TELOPHASE
46 MITOTICPHASE
47 NUCLEOSOME
2 GENETICRECOMBINATION
3 CELLDIVISION
5 CELLPLATE
6 ZYGOTE
7 CHROMATIN
9 DIPLOID
12 CLEAVAGEFURROW
14 CENTROSOME
16 CHEMOTHERAPY
18 GAMETE
22 MEIOSIS
25 SISTERCHROMATID
26 RADIATION
28 HOMOLOGOUS
31 METAPHASE
34 NONDISJUNCTION
40 MITOSIS
43 SEXUAL
44 SEX

Chapter 9

Content Quiz Answers

1. D
2. True
3. genetics
4. A
5. False, self-fertilize
6. F_2
7. A
8. A
9. D
10. False, heterozygous
11. locus
12. D
13. False, all
14. monohybrid
15. C
16. True
17. genotype
18. C
19. False, genotypes
20. D
21. B
22. False, are not
23. True
24. carriers
25. E
26. False, more
27. heterozygotes
28. Huntington's disease
29. cystic fibrosis
30. A
31. True
32. homozygous
33. B
34. False, did not show
35. hypercholesterolemia
36. A
37. C
38. False, dominant
39. AB
40. D
41. True
42. codominant
43. E
44. True
45. polygenic inheritance
46. A
47. True
48. genetic
49. D
50. B
51. C
52. True
53. False, linkage
54. chromosome theory of inheritance
55. fruit flies
56. D
57. A
58. True
59. False, occur
60. False, several
61. sex
62. hemophilia
63. False, artificial
64. wolves

Chapter 10

Content Quiz Answers

1. B
2. C
3. False, virus
4. B
5. True
6. structure
7. C
8. False, is missing an
9. deoxyribose, ribose
10. uracil
11. thymine/cytosine, adenine/guanine
12. polynucleotides, nucleotides
13. A
14. B
15. A
16. False, outside
17. GCTA
18. C
19. B
20. False, many replication origins
21. DNA polymerases
22. DNA polymerases
23. D
24. B
25. False, proteins
26. genotype, phenotype
27. transcription, translation
28. B
29. True
30. True
31. three
32. amino acids
33. A
34. C
35. False, no gaps
36. True
37. genetic code
38. C
39. True
40. promoter
41. RNA polymerase
42. A
43. True
44. introns, exons
45. B
46. E
47. D
48. False, anticodon
49. mRNA, tRNA
50. A
51. A
52. True
53. amino acid
54. A, P, translocation
55. stop codon
56. C
57. A
58. True
59. False, does not shift
60. False, not beneficial/harmful
61. mutation
62. mutagens
63. D
64. A
65. False, lytic
66. False, can
67. True
68. bacteriophage or phage
69. E
70. E
71. False, vaccines
72. True
73. mini-chromosomes
74. plasma membrane, nuclear membrane
75. B
76. plasma membrane
77. retrovirus
78. reverse transcriptase
79. D
80. True
81. RNA, protein
82. D
83. True
84. True
85. emerging

Chapter 11

Content Quiz Answers

1. A
2. True
3. gene
4. E
5. A
6. False, some
7. cellular differentiation
8. D
9. A
10. B
11. False, promoter
12. operon
13. D
14. C
15. B
16. D
17. D
18. B
19. False, prevent
20. True
21. False, DNA
22. enhancers
23. silencers
24. E
25. A
26. False, similar
27. True
28. False, cDNA
29. signal transduction
30. genes
31. hormones
32. A
33. D
34. True
35. True
36. False, nucleus
37. regeneration
38. False, 1950s
39. E
40. True
41. therapeutic
42. A
43. C
44. D
45. True
46. False, DNA
47. True
48. True
49. growth factors
50. E
51. B
52. False, fiber
53. carcinogens
54. D
55. True
56. evolutionary

Chapter 12

Content Quiz Answers

1. A
2. False, DNA
3. insecticide
4. A
5. False, bread
6. True
7. DNA technology
8. genetic engineering
9. E
10. True
11. False, genes
12. vitamin A
13. bacteria
14. vaccination
15. transgenic
16. milk
17. 4
18. 2
19. 7
20. 1
21. 6
22. 3
23. 5
24. D
25. False, separate from
26. True
27. nucleic acid probe
28. restriction
29. genomic library
30. vectors
31. bacteria
32. DNA ligase
33. C
34. True
35. DNA profiling
36. A
37. polymerase chain reaction (PCR)
38. short tandem repeat (STR)
39. DNA polymerase
40. repetitive
41. D
42. A
43. False, chromosomes
44. genomics
45. repetitive DNA
46. D
47. False, 96%
48. B
49. D
50. False, have not
51. False, many humans
52. genetic makeup
53. A
54. True
55. proteomics
56. C
57. True
58. marrow
59. E
60. True
61. C
62. True
63. B
64. True

1 STR
2 RESTRICTIONENZYME
3 NUCLEICACIDPROBE
4 VACCINE
5 HUMANGENETHERAPY
6 REPETITIVEDNA
7 TRANSGENIC
8 RECOMBINANTDNA
9 VECTOR
10 GENETICMARKER
11 GMORGANISM
12 GENOMICS
13 PROTEOMICS
14 RESTRICTIONFRAGMENT
15 PCR
16 GENOMICLIBRARY
17 HUMANGENOMEPROJECT
18 PLASMID
19 DNATECHNOLOGY
20 GENETICENGINEERING
21 DNAPROFILE
22 CLONE
23 DNALIGASE
24 BIOTECHNOLOGY
25 FORENSICS
26 GELELECTROPHORESIS
27 GENECLONING

Chapter 13

Content Quiz Answers

1. D
2. C
3. False, increase
4. False, selects for/favors
5. C
6. D
7. B
8. B
9. D
10. False, was not
11. True
12. True
13. gradualism
14. descent
15. A
16. False, prokaryotes
17. 3.5
18. D
19. C
20. A
21. False, does not support
22. A
23. E
24. True
25. common ancestor
26. B
27. D
28. True
29. False, most
30. South America
31. environment
32. A
33. True
34. True
35. C
36. homologous
37. ancestral
38. C
39. False, can
40. False, populations
41. gene pool
42. modern synthesis
43. A
44. False, phenotype
45. polygenic
46. C
47. D
48. A
49. B
50. B
51. D
52. E
53. C
54. A
55. C
56. B
57. Hardy-Weinberg
58. evolution or microevolution
59. D
60. A
61. D
62. D
63. False, decrease
64. True
65. True
66. genetic drift
67. E
68. A
69. C
70. B
71. C, D
72. A
73. True
74. stabilizing
75. extinction
76. D
77. dimorphism
78. C
79. A
80. True
81. malaria

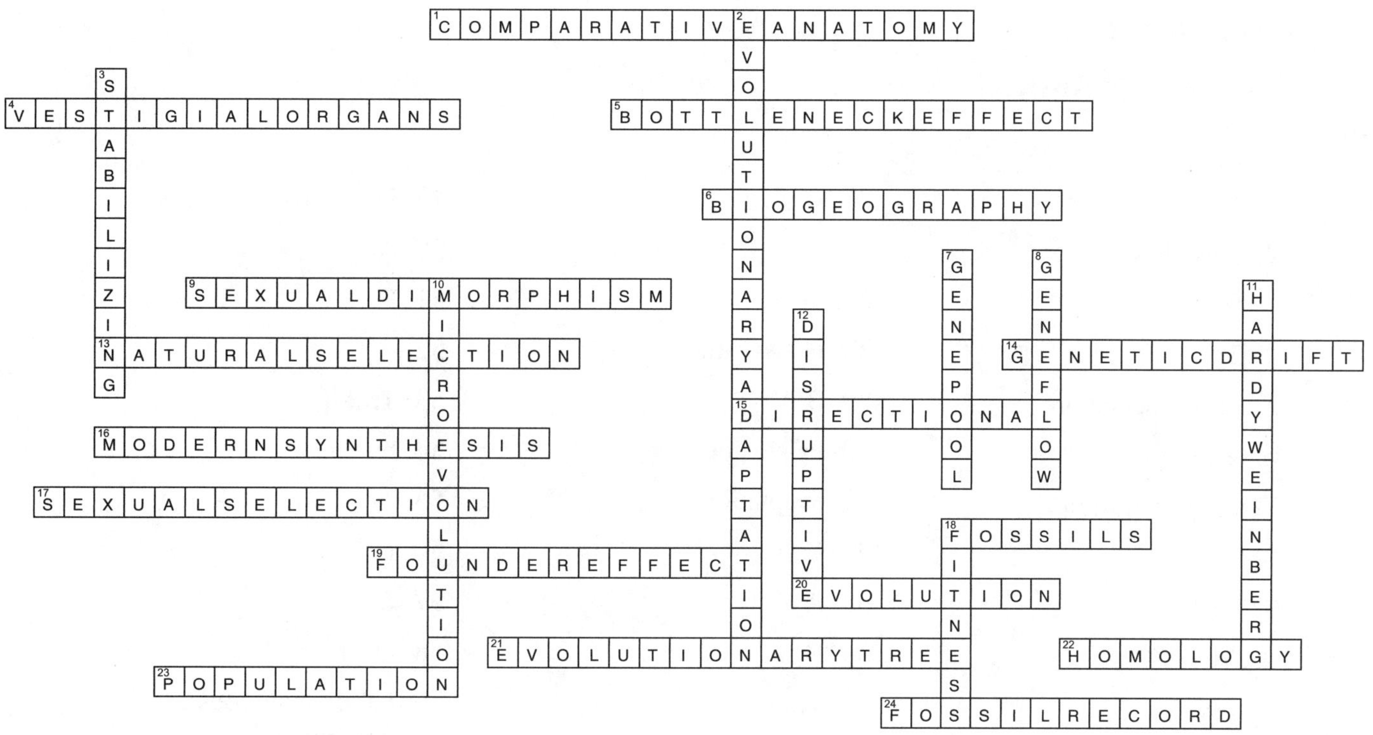
1 COMPARATIVEANATOMY
2 EVOLUTIONARYADAPTATION
3 STABILIZING
4 VESTIGIALORGANS
5 BOTTLENECKEFFECT
6 BIOGEOGRAPHY
7 GENEPOOL
8 GENEFLOW
9 SEXUALDIMORPHISM
10 MICROEVOLUTION
11 HARDYWEINBERG
12 DISRUPTIVE
13 NATURALSELECTION
14 GENETICDRIFT
15 DIRECTIONAL
16 MODERNSYNTHESIS
17 SEXUALSELECTION
18 FOSSILS
18 FITNESS
19 FOUNDEREFFECT
20 EVOLUTION
21 EVOLUTIONARYTREE
22 HOMOLOGY
23 POPULATION
24 FOSSILRECORD

Chapter 14

Content Quiz Answers

1. A
2. False, has not
3. extinction
4. E
5. False, speciation
6. Galápagos
7. macroevolution
8. D
9. True
10. True
11. fossil
12. D
13. C
14. True
15. hybrid sterility
16. D
17. D
18. False, two
19. sympatric
20. A
21. S
22. A
23. A
24. S
25. A
26. C
27. True
28. False, little
29. B
30. False, not have
31. exaptation
32. C
33. A
34. D
35. True
36. paedomorphosis
37. C
38. D
39. B
40. D
41. False, relative
42. radiometric dating
43. C
44. A
45. False, allopatric
46. continental drift
47. D
48. D
49. False, five
50. A
51. E
52. A
53. True
54. binomial
55. A
56. D
57. E
58. D
59. False, analogous, homologous
60. the same structure
61. analogous
62. homologous
63. convergent, homology
64. cladistics
65. E
66. False, Protista
67. cladistics
68. D
69. True
70. niches

1 PHYLOGENETIC TREE
2 PHYLUM
3 KINGDOM
4 PREZYGOTIC
5 BIOLOGICAL SPECIES
6 DOMAIN
7 THREE DOMAIN
8 CLASS
9 FAMILY
10 ALLOPATRIC
11 GEOLOGICAL TIME SCALE
12 CLADISTICS
13 ANALOGY
14 EVO DEVO
15 CLADE
16 MACROEVOLUTION
17 GENUS
18 PAEDOMORPHOSIS
19 CONVERGENT
20 BINOMIAL
21 SPECIES
22 RADIOMETRIC DATING
23 ORDER
24 SPECIATION
25 POSTZYGOTIC
26 PUNCTUATED EQUILIBRIUM
27 SYMPATRIC
28 SYSTEMATICS
29 TAXONOMY
30 REPRODUCTIVE BARRIER

Chapter 15

Content Quiz Answers

1. D
2. False, has not
3. genome-free
4. A
5. B
6. D
7. False, prokaryotes
8. True
9. cellular respiration
10. E
11. False, different
12. spontaneous generation
13. D
14. B
15. True
16. False, would have
17. ribozymes/other RNA
18. E
19. False, could
20. True
21. bacteria
22. F
23. C
24. E
25. A
26. B
27. D
28. B
29. C
30. D
31. A
32. D
33. A
34. C
35. True
36. binary fission
37. photoautotrophs
38. chemoheterotrophs
39. C
40. False, anaerobic
41. archaea
42. D
43. B
44. E
45. False, are not
46. exotoxins
47. B
48. bioremediation
49. B
50. E
51. C
52. C
53. False, aerobic
54. mitochondria
55. protists
56. H
57. K
58. G
59. I
60. D
61. B
62. F
63. E
64. J
65. A
66. C
67. G
68. D
69. E
70. B
71. F
72. C
73. A
74. D
75. B
76. D
77. B
78. True
79. green algae
80. pigments
81. D
82. True
83. *Volvox*

1 PROTIST
2 PATHOGEN
3 CELLULAR SLIME MOLD
4 PLANKTON
1 PROKARYOTE
5 CILIATE
6 PROTOZOA
7 FLAGELLATES
8 EUKARYOTES
9 SEAWEED
10 RIBOZYME
11 BACILLI
12 APICOMPLEXA
13 SYMBIOSIS
14 ARCHAEA
15 DINOFLAGELLATE
16 BACTERIA
17 BIOREMEDIATION
18 AMOEBA
19 BINARY FISSION
20 ALGA
21 SPONTANEOUS GENERATION
22 ENDOSYMBIOSIS
23 DIATOM
24 GREEN ALGAE
25 EXOTOXIN
26 COCCI
27 ENDOSPORE
28 BIOGENESIS
29 ENDOTOXIN
30 PSEUDOPODIA
31 PLASMODIAL SLIME MOLD

Chapter 16

Content Quiz Answers

1. C
2. True
3. symbiosis
4. C
5. E
6. D
7. A
8. A
9. B
10. A
11. D
12. True
13. mycorrhizae
14. stomata
15. cuticle
16. roots, shoots
17. lignin
18. A
19. True
20. charophytes
21. B
22. D
23. C
24. A
25. False, angiosperms
26. seed
27. D
28. B
29. False, diploid
30. False, gametophyte
31. sporophyte, gametophyte
32. alternation of generations
33. B
34. True
35. coal
36. C
37. E
38. A
39. C
40. True
41. False, do not have
42. ovules
43. pollen
44. female gametophytes
45. D
46. F
47. B
48. E
49. C
50. A
51. A
52. A
53. True
54. False, angiosperms
55. False, flower
56. fruit
57. plants
58. C
59. True
60. True
61. C
62. A
63. D
64. False, chitin
65. spores
66. enzymes
67. hyphae
68. mycelium
69. H
70. False, less
71. decomposers
72. ergot
73. D
74. False, symbiotic
75. fungi
76. lichen

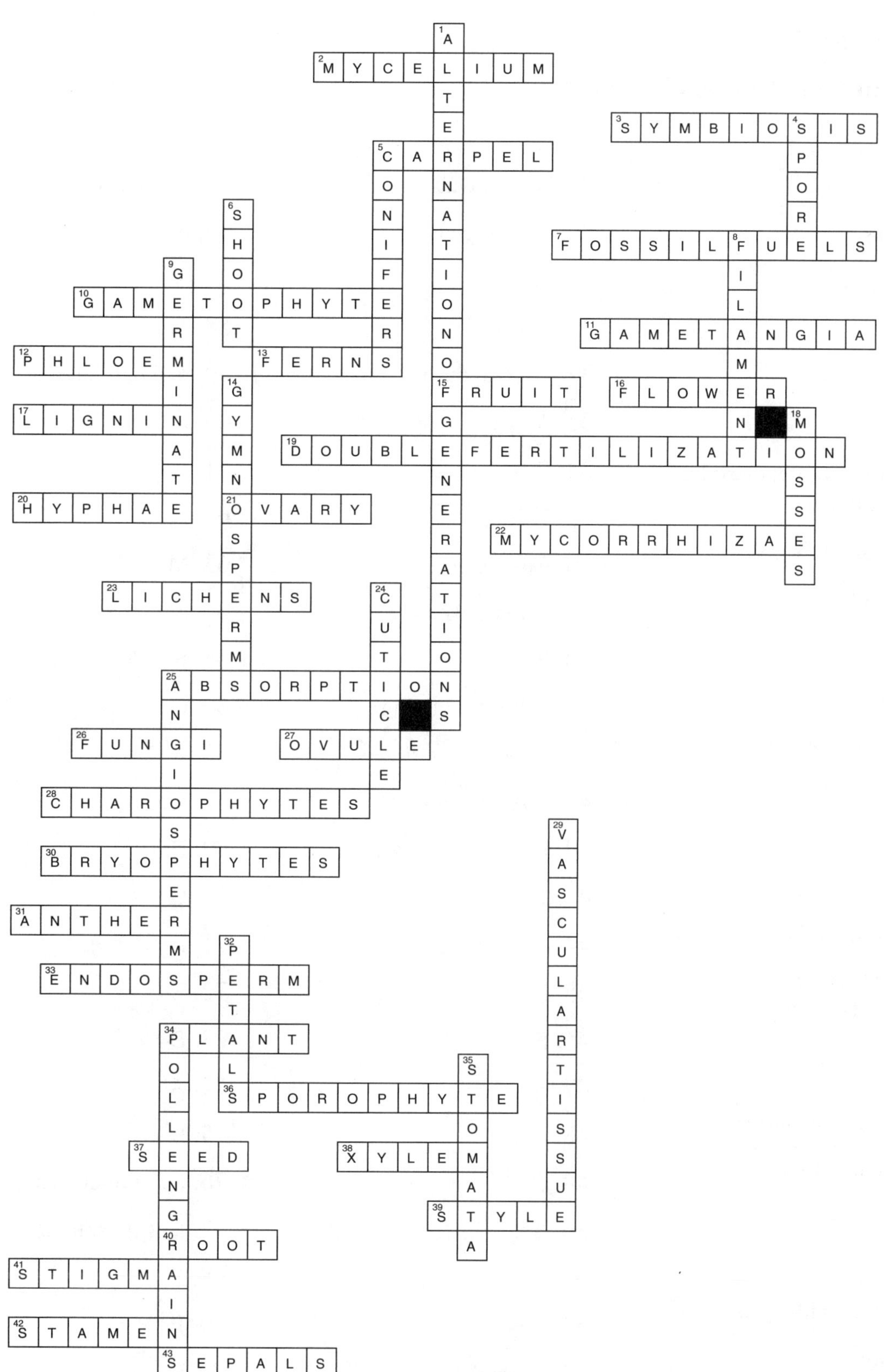

1 ALTERNATION OF GENERATIONS
2 MYCELIUM
3 SYMBIOSIS
4 SPORS
5 CARPEL
5 CONIFERS
6 SHOOT
7 FOSSIL FUELS
8 FILAMENT
9 GERMINATE
10 GAMETOPHYTE
11 GAMETANGIA
12 PHLOEM
13 FERNS
14 GYMNOSPERM
15 FRUIT
16 FLOWER
17 LIGNIN
18 MOSSES
19 DOUBLE FERTILIZATION
20 HYPHAE
21 OVARY
22 MYCORRHIZAE
23 LICHENS
24 CUTICLE
25 ABSORPTION
25 ANGIOSPERM
26 FUNGI
27 OVULE
28 CHAROPHYTES
29 VASCULAR TISSUE
30 BRYOPHYTES
31 ANTHER
32 PETAL
33 ENDOSPERM
34 PLANT
34 POLLEN GRAIN
35 STOMATA
36 SPOROPHYTE
37 SEED
38 XYLEM
39 STYLE
40 ROOT
41 STIGMA
42 STAMEN
43 SEPALS

Chapter 17

Content Quiz Answers

1. A
2. True
3. islands
4. D
5. A
6. True
7. False, zygote, blastula, gastrula
8. metamorphosis
9. muscle, nerve
10. C
11. True
12. False, modern
13. E
14. C
15. A
16. False, bilateral
17. pseudocoelom
18. D
19. D
20. False, are marine
21. choanocytes
22. B
23. A
24. False, cnidocytes
25. polyp
26. medusa
27. D
28. False, bivalves
29. cephalopods
30. radula
31. mantle
32. visceral mass
33. A
34. C
35. True
36. False, can
37. blood flukes
38. B
39. False, free-living
40. True
41. soil, castings
42. B
43. True
44. decomposers
45. D
46. E
47. B
48. A
49. C
50. B
51. D
52. True
53. exoskeleton
54. crustaceans
55. segments
56. insects
57. A
58. True
59. radial, bilateral
60. L
61. J
62. F
63. M
64. N
65. B
66. C
67. A
68. K
69. H
70. G
71. I
72. E
73. D
74. A
75. B
76. False, includes
77. lancelets, tunicates
78. F
79. B
80. D
81. E

82. A
83. C
84. D
85. True
86. lateral line
87. operculum
88. swim bladder
89. C
90. D
91. E
92. False, tetrapods
93. skin
94. A
95. E
96. B
97. D
98. True
99. False, fewer
100. lungs
101. amniotic
102. endotherms
103. A
104. D
105. False, marsupial
106. monotremes
107. A
108. A
109. True
110. False, Old World
111. anthropoids
112. arboreal
113. C
114. E
115. D
116. A
117. B
118. A
119. C
120. E
121. A
122. False, are not
123. False, at different rates
124. True
125. True
126. 4
127. culture
128. C
129. True

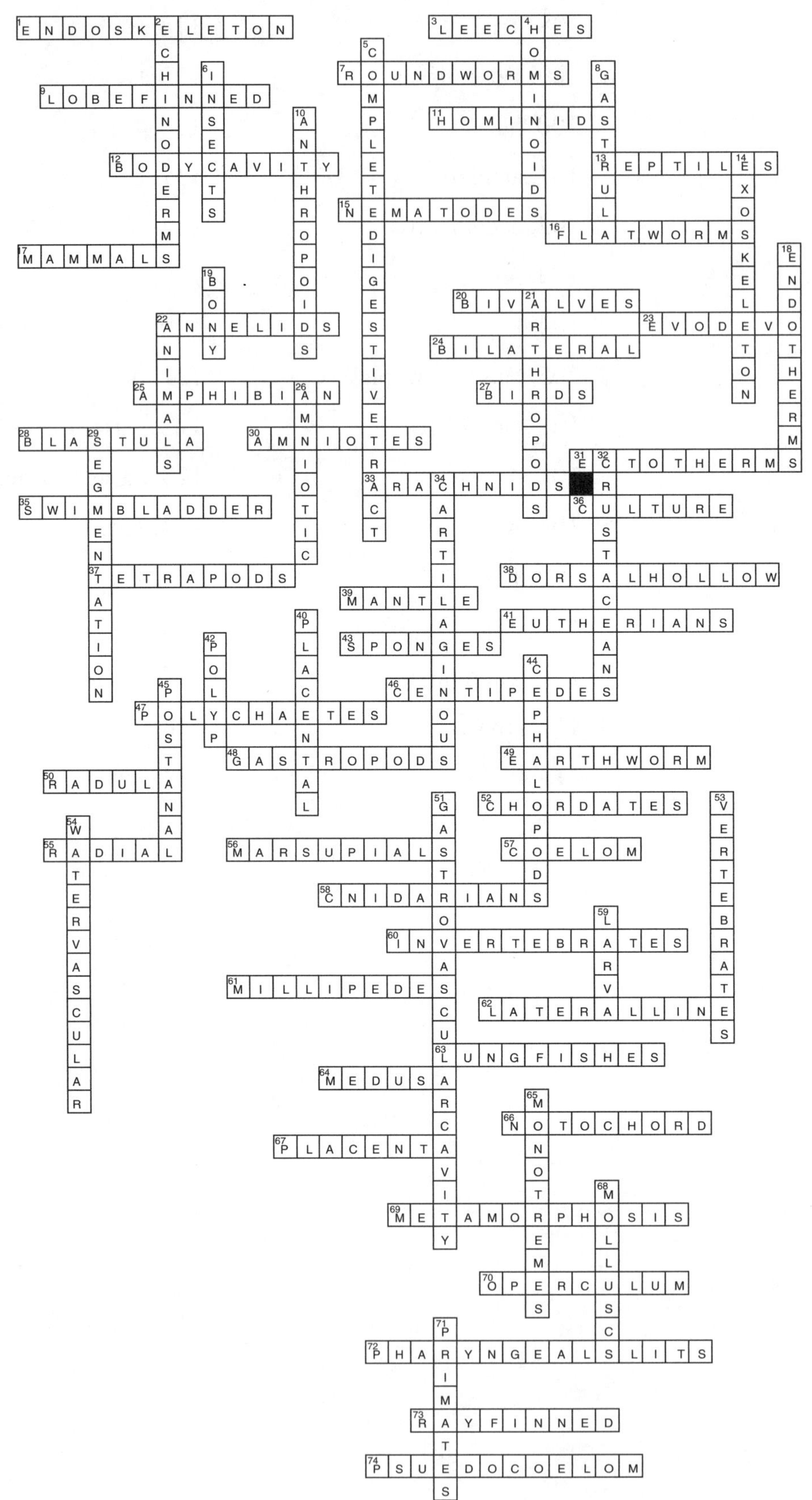

Chapter 18

Content Quiz Answers

1. E
2. True
3. human
4. A
5. True
6. discovery-based
7. B
8. C
9. A
10. D
11. E
12. True
13. biosphere
14. abiotic
15. ecology
16. habitat
17. B
18. A
19. D
20. True
21. False, enzymes
22. False, top/surface
23. False, seldom
24. nitrogen and phosphorus
25. chemoautotrophic
26. C
27. A
28. B
29. D
30. True
31. False, plants, animals
32. acclimation
33. anatomical
34. D
35. C
36. A
37. B
38. E
39. C
40. B
41. False, high
42. True
43. B
44. B
45. True
46. True
47. False, phytoplankton
48. False, photic
49. zooplankton
50. estuary
51. pelagic
52. aphotic
53. hydrothermal vent
54. A
55. D
56. True
57. rain forests
58. temperate
59. heat
60. B
61. D
62. C
63. G
64. H
65. F
66. E
67. J
68. A
69. I
70. E
71. D
72. A
73. B
74. C
75. A
76. True
77. climograph
78. convergent
79. D
80. True
81. water cycle
82. E
83. False, agriculture
84. sustainability
85. freshwater
86. B
87. True
88. greenhouse effect
89. D
90. False, carbon dioxide
91. True
92. acidic
93. become extinct

94. E
95. False, northward
96. daylength
97. B
98. True
99. landfills
100. carbon
101. C
102. True
103. extinction

1 ECOLOGY
2 POPULATION
3 TROPICALFORESTS
4 TUNDRA
5 SUSTAINABILITY
6 POPULATION
7 ZOOPLANKTON
8 PERMAFROST
9 WETLAND
10 POLARICE
11 GREENHOUSEEFFECT
12 GREENHOUSEGASES
13 INTERTIDAL
14 TEMPERATE
15 HABITAT
16 PHYTOPLANKTON
17 SAVANNAS
18 TAIGA
19 PHOTIC
20 PELAGIC
21 ORGANISMAL
22 CONIFEROUSFORESTS
23 DESERT
24 BIOTIC
25 ECOSYSTEM
26 CARBONFOOTPRINT
27 CORALREEF
28 CHAPARRAL
29 ACCLIMATION
30 ABIOTIC
31 COMMUNITY
32 BIOME
33 COMMUNITY
34 TROPICS
35 APHOTIC
36 ESTUARY
37 BIOSPHERE
38 ECOSYSTEM

Chapter 19

Content Quiz Answers

1. E
2. True
3. non-native/invasive
4. D
5. False, population
6. population ecology
7. applied
8. B
9. C
10. D
11. A
12. B
13. C
14. True
15. individuals
16. A
17. True
18. life table
19. survivorship curve
20. II
21. C
22. A
23. True
24. opportunistic
25. life history
26. False, little/no
27. B
28. B
29. False, decreases
30. True
31. True
32. limiting
33. exponential
34. resources
35. C
36. D
37. D
38. E
39. False, increases
40. True
41. True
42. density-dependent
43. intraspecific competition
44. B
45. A
46. True
47. False, slower
48. endangered
49. threatened
50. C
51. False, opportunistic
52. invasive
53. C
54. D
55. False, viruses
56. biological control
57. E
58. True
59. monoculture
60. pests
61. A
62. C
63. False, a decrease
64. birth, death
65. C
66. A
67. True
68. age structure
69. population momentum
70. E
71. False, overconsumption
72. True
73. ecological footprint
74. E
75. True
76. invasive

Chapter 20

Content Quiz Answers

1. B
2. True
3. plants
4. E
5. D
6. True
7. diversity/variation
8. services
9. C
10. A
11. True
12. invasive
13. D
14. C
15. C
16. D
17. B
18. C
19. B
20. False, cryptic
21. True
22. environment
23. community
24. interspecific
25. interspecific competition
26. niche
27. warning
28. C
29. D
30. A
31. B
32. B
33. B
34. D
35. C
36. E
37. False, producer
38. trophic levels
39. food chain
40. prokaryotes/fungi
41. webs
42. A
43. C
44. B
45. True
46. True
47. keystone
48. ecological succession
49. A
50. D
51. E
52. False, sunlight
53. True
54. True
55. True
56. True
57. ecosystem
58. biomass
59. primary productivity
60. plants
61. abiotic
62. air/soil
63. A
64. C
65. False, nitrogen
66. True
67. True
68. biogeochemical
69. phosphate
70. nitrogen
71. abiotic
72. carbon
73. A
74. D
75. True
76. endemic
77. E
78. D
79. False, landscape
80. landscape ecology
81. movement corridor
82. E
83. True
84. bioremediation
85. D
86. True
87. biodiversity
88. biophilia

Chapter 21

Content Quiz Answers

1. A
2. False, exhaustion
3. heat stroke
4. A
5. True
6. anatomy, physiology
7. D
8. A
9. B
10. C
11. B
12. A
13. False, neuron
14. loose connective tissue
15. B
16. False, organ
17. organ systems
18. B
19. False, open
20. circulatory
21. D
22. True
23. True
24. homeostasis
25. A
26. A
27. False, inhibit
28. positive
29. D
30. B
31. False, endotherms
32. thermoregulation
33. E
34. True
35. False, osmoregulators
36. osmoregulation
37. C
38. A
39. B
40. D
41. A
42. A
43. D
44. True
45. dialysis
46. D
47. False, raising
48. behavioral
49. anatomical
50. physiological

Chapter 22

Content Quiz Answers

1. B
2. True
3. obese
4. D
5. C
6. A
7. B
8. B
9. C
10. A
11. False, Chemical
12. monomers
13. food vacuoles
14. I
15. F
16. A
17. H
18. K
19. J
20. C
21. L
22. B
23. E
24. G
25. D
26. N
27. M
28. D
29. C
30. False, heartburn
31. amylase
32. peristalsis
33. sphincters
34. rectum
35. A
36. True
37. False, plant
38. True
39. False, inorganic
40. basal metabolic rate
41. diet
42. enzymes
43. minerals
44. energy
45. heat
46. D
47. False, Bulimia
48. True
49. protein
50. obese
51. B
52. True
53. fatty

Chapter 23

Content Quiz Answers

1. B
2. True
3. circulatory
4. C
5. B
6. False, a closed
7. False, carbon dioxide
8. cardiovascular
9. D
10. True
11. double
12. A
13. C
14. True
15. False, speed up
16. diastole
17. heart murmur
18. AED
19. C
20. A
21. True
22. True
23. hypertension
24. E
25. A
26. C
27. False, leukocytes
28. True
29. plasma
30. hemoglobin
31. B
32. False, heart attack
33. False, heart attack
34. interstitial fluid
35. atherosclerosis
36. B
37. False, more
38. a tracheal system
39. gills
40. oxygen, ATP
41. A
42. B
43. False, alveolus
44. vocal cords
45. B
46. False, diaphragm
47. carbon dioxide
48. negative
49. C
50. D
51. True
52. smoking/not smoking
53. D
54. False, pharynx
55. digestive

1 PULMONARY
2 NEGATIVE PRESSURE
3 SYSTOLE
4 SYSTEMIC
5 SA
1 PLATELETS
7 PULSE
8 RESPIRATORY
9 PHARYNX
10 RESPIRATORY
6 TRACHEAL SYSTEM
11 ERYTHROCYTE
12 FIBRINOGEN
13 PACEMAKER
14 HEART
15 VEINS
16 CARDIOVASCULAR
17 HEMOGLOBIN
18 CORONARY
19 DIAPHRAGM
20 GILLS
21 CLOSED
22 ATRIOVENTRICULAR
23 INTERSTITIAL FLUID
24 HEART ATTACK
25 LEUKOCYTES
26 CIRCULATORY
27 DIFFUSION
28 DIASTOLE
29 HYPERTENSION
30 CARDIOVASCULAR
31 ANEMIA
32 LUNG
33 ALVEOLI
33 ARTERIES
34 FIBRIN
35 CARDIAC
36 ATHEROSCLEROSIS
37 LARYNX
37 LEUKEMIA
38 BRONCHI
39 CAPILLARIES
40 ARTERIOLES
41 HEART MURMUR
42 TRACHEA
43 BREATHING
44 OPEN
45 BLOOD PRESSURE
46 BRONCHIOLES
47 WHITE
48 DOUBLE
49 ATRIUM
50 VENULES
51 VENTRICLE
52 PLASMA
53 RED
54 HEART RATE
55 VOCAL CORDS

Chapter 24

Content Quiz Answers

1. A
2. False, has not
3. True
4. vaccine
5. D
6. True
7. mucus
8. antimicrobial chemicals/enzymes
9. C
10. True
11. complement
12. phagocytic
13. E
14. A
15. False, but do not address
16. damaged
17. C
18. False, lymph
19. lymph nodes
20. E
21. False, adaptive
22. False, lymphocytes
23. T cells/T lymphocytes
24. antigen
25. B
26. D
27. True
28. False, antigen
29. True
30. Y
31. monoclonal antibody
32. HIV
33. D
34. A
35. E
36. C
37. D
38. A
39. False, ineffective
40. True
41. cell-mediated
42. humoral
43. memory
44. antibodies
45. vaccination or immunization
46. E
47. False, histamine
48. True
49. allergens
50. anaphylactic shock
51. B
52. D
53. C
54. A
55. C
56. True
57. autoimmune
58. A
59. True
60. lymphocytes
61. Immunodeficiency
62. immune
63. helper T
64. D
65. False, are
66. natural selection

Chapter 25

Content Quiz Answers

1. D
2. True
3. menopause
4. C
5. D
6. False, endocrine
7. steroid
8. hormone
9. target
10. endocrine
11. A
12. True
13. glucose
14. D
15. C
16. False, inhibits
17. FSH and LH
18. endorphins
19. dwarfism
20. E
21. B
22. E
23. True
24. goiter
25. antagonistic
26. calcitonin
27. D
28. True
29. glucagon
30. diabetes mellitus
31. insulin
32. sugar
33. D
34. E
35. False, medulla
36. glucocorticoids
37. A
38. D
39. True
40. progestins
41. estrogens
42. androgens
43. B
44. F
45. G
46. A
47. H
48. E
49. C
50. D
51. A
52. False, downstream
53. endocrine disruptors
54. B
55. False, early

1 PARATHYROID
2 POSTERIORPITUITARY
3 ANTAGONISTIC
4 PANCREAS
5 TARGET
6 GLUCOCORTICOID
7 THYROID
8 PROGESTINS
9 GLUCAGON
10 HORMONE
11 NOREPINEPHRINE
12 DIABETESMELLITUS
13 ENDOCRINE
14 EPINEPHRINE
15 ANTERIORPITUITARY
16 ADRENALCORTEX
17 ENDORPHIN
18 CORTICOSTEROIDS
19 ADRENALMEDULLA
20 GONAD
21 CALCITONIN
22 ANDROGENS
23 PTH
24 HYPOTHALAMUS
25 ADRENAL
26 ESTROGENS
27 GH
28 INSULIN
29 PITUITARY

Chapter 26

Content Quiz Answers

1. B
2. D
3. True
4. infertility
5. B
6. D
7. A
8. C
9. C
10. False, uniform
11. asexual
12. budding
13. regeneration
14. D
15. True
16. False, internal
17. sperm, ovum
18. zygote
19. hermaphrodites
20. E
21. C
22. F
23. A
24. D
25. B
26. G
27. A
28. False, below
29. ejaculation
30. A
31. D
32. F
33. B
34. H
35. G
36. E
37. C
38. E
39. True
40. gonads
41. ovulation
42. C
43. D
44. True
45. gametogenesis
46. primary oocytes
47. seminiferous tubules
48. epididymis
49. B
50. False, ovarian
51. False, FSH
52. menstruation
53. 14
54. anterior pituitary gland
55. HCG, human chorionic gonadotropin
56. D
57. B
58. F
59. A
60. G
61. C
62. H
63. E
64. D
65. True
66. RU-486 (mifepristone)
67. B
68. True
69. True
70. bacteria
71. fungi
72. viruses
73. E
74. False, haploid
75. acrosome
76. mitochondria
77. C
78. A
79. F
80. G
81. H
82. D
83. B
84. E
85. C
86. True
87. A
88. C
89. B
90. D
91. A
92. False, can

93. trophoblast
94. placenta
95. stem
96. D
97. False, second
98. third
99. first
100. first
101. C
102. False, prolactin
103. expulsion
104. labor
105. placenta
106. dilation
107. E
108. A
109. False, man
110. impotence
111. impotence
112. C
113. True
114. "in glass"
115. A
116. False, do not lose/retain
117. menopause

Chapter 27

Content Quiz Answers

1. D
2. False, electroreception
3. echolocation
4. A
5. False, motor
6. False, peripheral
7. nerve
8. sensory input
9. cell body
10. supporting
11. myelin sheath
12. axons
13. sensory receptors
14. B
15. C
16. C
17. D
18. True
19. False, negatively
20. resting potential
21. stimulus
22. threshold
23. F
24. E
25. C
26. H
27. I
28. B
29. A
30. D
31. G
32. B
33. D
34. True
35. False, an electrical
36. synaptic cleft
37. synapse
38. chemical
39. neurotransmitter
40. D
41. True
42. False, central
43. False, paraplegia
44. bacterial
45. brain
46. meningitis
47. D
48. False, somatic
49. False, parasympathetic
50. False, somatic
51. sympathetic
52. G
53. C
54. B
55. E
56. F
57. A
58. D
59. B
60. False, cerebellum
61. lateralization
62. Alzheimer's
63. E
64. D
65. A
66. B
67. C
68. B
69. False, transduction
70. True
71. receptor potential
72. prostaglandins
73. F
74. M
75. J
76. O
77. P
78. L
79. E
80. C
81. N
82. A
83. I
84. B
85. K
86. G
87. D

88. H
89. C
90. False, aqueous
91. rhodopsin, photopsin
92. B
93. A
94. E
95. C
96. D
97. F
98. 8
99. 5
100. 7
101. 1
102. 3
103. 9
104. 2
105. 6
106. 4
107. C
108. E
109. False, middle
110. tinnitus
111. C
112. A
113. False, appendicular
114. True
115. False, yellow
116. pivot
117. ball-and-socket
118. hinge
119. A
120. B
121. D
122. A
123. D
124. True
125. False, Thick
126. True
127. tendons
128. sarcomeres
129. myofibrils
130. motor unit
131. neuromuscular
132. E
133. True

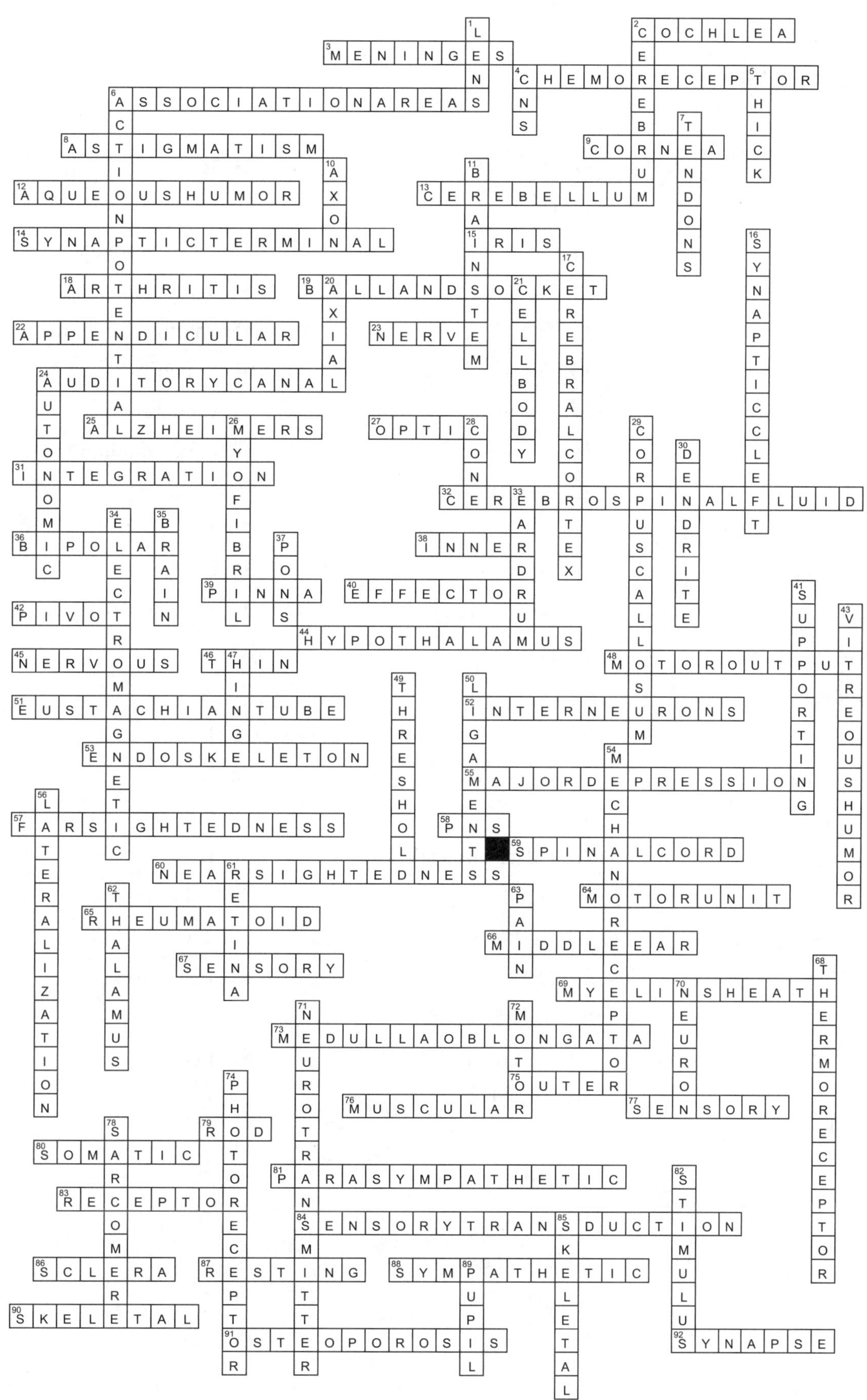

Chapter 28

Content Quiz Answers

1. E
2. True
3. False, plants
4. green
5. C
6. D
7. False, two leaves
8. cotyledons
9. A
10. B
11. False, stems
12. True
13. False, leaves
14. taproots
15. shoot
16. nodes, internodes
17. apical dominance
18. axillary
19. runner
20. root, stem (called a rhizome)
21. roots
22. root hairs
23. leaves
24. tendril
25. I
26. E
27. A
28. B
29. J
30. F
31. D
32. H
33. C
34. G
35. True
36. True
37. vascular
38. ground
39. dermal
40. F
41. D
42. B
43. H
44. A
45. E
46. C
47. G
48. False, cell wall
49. cellulose
50. tracheids, vessel elements
51. D
52. False, second
53. perennials
54. annuals
55. B
56. False, primary
57. root cap
58. apical meristems
59. primary
60. A
61. B
62. D
63. False, secondary
64. True
65. vascular cambium
66. xylem
67. drought
68. cork cambium
69. bark
70. E
71. C
72. F
73. H
74. A
75. I
76. D
77. G
78. B
79. False, leaves
80. True
81. False, asexual
82. flower
83. pistil
84. D
85. A
86. C
87. B
88. E
89. F
90. False, stigma

91. double fertilization
92. endosperm
93. D
94. seed coat
95. endosperm
96. D
97. E
98. False, root
99. True
100. seeds
101. C
102. True
103. ultraviolet

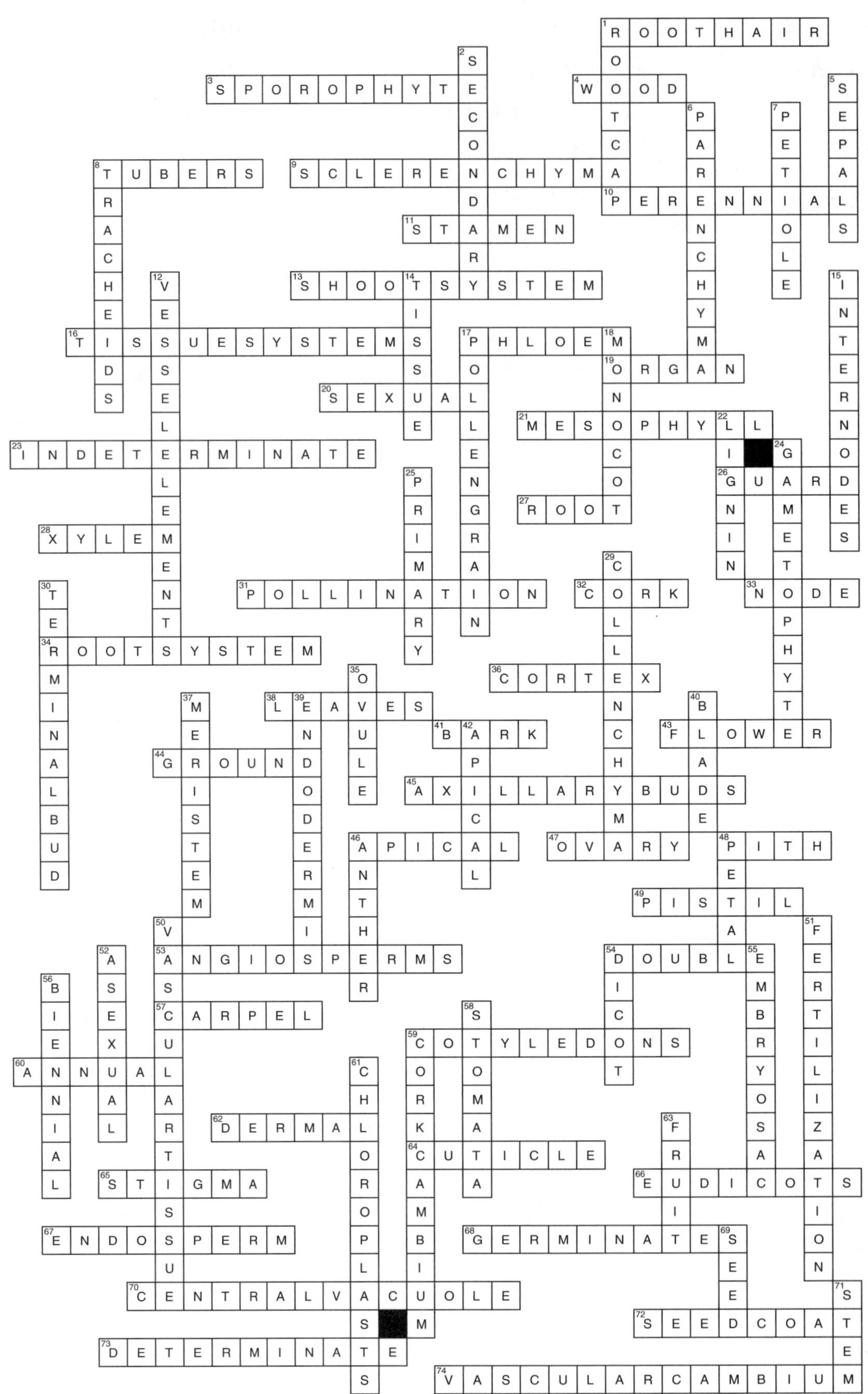

Chapter 29

Content Quiz Answers

1. D
2. True
3. time
4. E
5. C
6. B
7. False, macronutrients
8. True
9. organic
10. essential
11. compost
12. fertilizer
13. nitrogen
14. B
15. True
16. root hairs
17. mycorrhiza
18. B
19. A
20. E
21. C
22. F
23. D
24. False, cannot
25. nitrogen
26. A
27. C
28. B
29. False, cohesion
30. True
31. xylem
32. C
33. False, source
34. True
35. phloem
36. B
37. C
38. E
39. A
40. D
41. A
42. D
43. D
44. A
45. False, faster
46. True
47. True
48. False, carbon dioxide
49. hormone
50. auxins
51. phototropism
52. C
53. False, spring or early summer
54. thigmotropism
55. gravitropism
56. photoperiod
57. E
58. True
59. algae and plants
60. insects

1 TRANSPIRATIONCOHESIONTENSION
2 GRAVITROPISM
3 GIBBERELLIN
4 MACRONUTRIENTS
5 TROPISM
6 MYCORRHIZA
7 MICRONUTRIENTS
8 HORMONES
9 ORGANIC
10 ABSCISICACID
11 ADHESION
12 CYTOKININS
13 ESSENTIAL
13 ETHYLENE
14 COMPOST
15 COHESION
16 PHLOEMSAP
17 XYLEMSAP
18 NITROGENFIXATION
19 MINERALS
20 FERTILIZER
21 THIGMOTROPISM
22 PHOTOTROPISM
23 PRESSUREFLOW
24 SUGARSOURCE
25 TRANSPIRATION
26 SUGARSINK
27 PHOTOPERIOD